Oxford Series in Ecology and Evolution

Edited by Paul H. Harvey, Robert M. May, H. Charles J. Godfr. ... ... Dunne

The Ecology of Adaptive Radiation
Dolph Schluter

Parasites and the Behavior of Animals
Janice Moore

Evolutionary Ecology of Birds
Peter Bennett and Ian Owens

The Role of Chromosomal Change in Plant Evolution
Donald A. Levin

Living in Groups
Jens Krause and Graeme D. Ruxton

Stochastic Population Dynamics in Ecology and Conservation
Russell Lande, Steiner Engen, and Bernt-Erik Sæther

The Structure and Dynamics of Geographic Ranges
Kevin J. Gaston

Animal Signals
John Maynard Smith and David Harper

Evolutionary Ecology: The Trinidadian Guppy
Anne E. Magurran

Infectious Diseases in Primates: Behavior, Ecology, and Evolution
Charles L. Nunn and Sonia Altizer

Computational Molecular Evolution
Ziheng Yang

The Evolution and Emergence of RNA Viruses
Edward C. Holmes

Aboveground–Belowground Linkages: Biotic Interactions, Ecosystem Processes, and Global Change
Richard D. Bardgett and David A. Wardle

Principles of Social Evolution
Andrew F. G. Bourke

Maximum Entropy and Ecology: A Theory of Abundance, Distribution, and Energetics
John Harte

Ecological Speciation
Patrik Nosil

Energetic Food Webs: An Analysis of Real and Model Ecosystems
John C. Moore and Peter C. de Ruiter

Evolutionary Biomechanics: Selection, Phylogeny, and Constraint
Graham K. Taylor and Adrian L. R. Thomas

Evolutionary Biomechanics

Selection, Phylogeny, and Constraint

GRAHAM K. TAYLOR

University of Oxford, UK

ADRIAN L. R. THOMAS

University of Oxford, UK

OXFORD
UNIVERSITY PRESS

OXFORD

UNIVERSITY PRESS

Great Clarendon Street, Oxford, OX2 6DP,
United Kingdom

Oxford University Press is a department of the University of Oxford.
It furthers the University's objective of excellence in research, scholarship,
and education by publishing worldwide. Oxford is a registered trade mark of
Oxford University Press in the UK and in certain other countries

First Edition published in 2014
Reprinted 2014

Impression: 2

Published in the United States of America by Oxford University Press
198 Madison Avenue, New York, NY 10016, United States of America

British Library Cataloguing in Publication Data
Data available

Library of Congress Control Number: 2013945536

ISBN 978-0-19-856637-3 (hbk.)
ISBN 978-0-19-856638-0 (pbk.)

Printed in Great Britain by
Clays Ltd, St Ives plc

To Our Parents

Preface

This book grew out of a suggestion from Paul Harvey, to whom we are greatly indebted. Like the biomechanical systems that it discusses, this book has evolved over an extended, if not quite geological, period of time. It has, in fact, outlived several generations of assistant editors at Oxford University Press, and we thank Ian Sherman for his forbearance as Senior Commissioning Editor throughout. We thank Lucy Nash for helping us see the project to completion, and our research groups for their patience as our time was absorbed to this end. We would also like particularly to thank our own families—Rachel, Isabelle and Isaac—Sue and Lauren—for their patience and support throughout the process.

In our defence, if we have taken our time in writing this book, it is because the material that it contains is almost entirely new. It was, after all, only a decade ago that Lauder (2003) was able to write: '*The discipline of biomechanics has had a long but relatively superficial flirtation with evolutionary biology. ... biomechanics and the allied areas of functional morphology and comparative physiology have focused largely on the question of how organisms work ... only to a lesser extent have issues relating to historical patterns and transformations been current within biomechanical research programmes*'. Lauder went on to propose a series of challenges for integrating evolutionary thinking into biomechanics—any one of which could form the basis of a research career in its own right. We would go a step further, however, and argue that the greatest challenge lies not in integrating evolutionary biology into biomechanics, but in integrating biomechanics into evolutionary biology.

In our view, evolutionary biomechanics should not be a niche specialism aimed only at studying the evolution of biomechanical systems. Rather, evolutionary biomechanics should be a discipline of much broader interest, aimed at understanding evolution *through* the study of biomechanical systems. Biomechanics is uniquely well placed to contribute to analyses of evolution because of the precision with which it is possible to predict physical constraints and physical performance from first principles. This precision can bring into sharp focus key questions about evolution which would otherwise remain vague or ill posed. We see a great opportunity here, waiting to be grasped. For example, much as evo-devo has helped to make sense of developmental constraints and possibilities in evolution, so evolutionary biomechanics can help to make sense of physical constraints and possibilities, and their interactions with the biological process of natural selection.

Theoretical evolutionary biology is founded upon the mathematics of population genetics; theoretical biomechanics expresses the mathematics of Newtonian mechanics; and comparative biology is grounded upon the mathematics of statistical modelling. It is difficult to escape the conclusion that a book which aims to build the foundations of evolutionary biomechanics must—by necessity—be mathematical through and through. We have not been shy of this, but have always tried to follow as our guiding principle that '*the supreme goal of all theory is to make the irreducible basic elements as simple and as few as possible without having to surrender the adequate representation of a single datum of experience*' (Einstein, 1934). We have taken the approach of verbally explaining all of the mathematics presented in the main text, with the goal of making it fully comprehensible to any reader with a solid grasp of high school mathematics. In consequence, the text is not easy-going in some places, but we have confined all of the more involved mathematics to text boxes, to avoid interrupting the flow.

Fundamentally, this is a book about evolutionary biology and the contribution that evolutionary biomechanics can make to that discipline. We make no attempt or pretence to be exhaustive in our coverage, and there are certainly many important areas of biomechanical investigation that do not receive attention here, including materials, structures, swimming and feeding. Biomechanics is already blessed with some superb textbooks and collected volumes, and for recent surveys we refer the reader to: Dudley (2000), Alexander (2003), Bels et al. (2003), Biewener (2003), Vogel (2003), Videler (2005), Pennycuick (2008), Taylor et al. (2010) and Ennos (2012). Hence, instead of attempting to review the biomechanical literature in a broad sense, we have chosen to explore a few examples in depth. These examples serve as vehicles for discussing fundamental concepts, analytical techniques and ideas about evolution. In summary, this is book that is about population genetics, optimization theory, dimensional analysis, statistical models and the comparative method as much as it is a book about biomechanics and the physics of life.

Numerous colleagues have contributed to the realisation of this project. Particular thanks are due to Paul Harvey for his comments upon many of the chapters, and to members of the Oxford Animal Flight Group, and Theresa Burt de Perera for feedback upon various parts of the book. Several cohorts of Biological Sciences undergraduates at Oxford have played an important part in pointing out where concepts were especially difficult to follow. We thank Alan Grafen for comments on a draft of Chapter 2, and Rafał Żbikowski for helpful advice on the mathematical content of that chapter. A conversation with Steve Frank on Fisher's Fundamental Theorem provided prompts on some points that would need clarification when Chapter 2 was written, and we thank him also for drawing our attention to some literature relevant to Chapter 8. Caroline Warman commented upon our translation of Galileo in Chapter 4.

Special thanks are due to R. McNeill Alexander, Young-Hui Chang and Max Donelan for allowing us access to original data: we hope they will be happy with the ways in which we have presented these data in Chapter 3. Ian Owens contributed hugely to an earlier version of the comparative dataset on bird flight morphology that we analyse in Chapters 5–8. Finally, we thank our colleagues in other disciplines with whom we have had the opportunity to interact through the unique environment of an Oxford college. Although they may not be aware of the fact, lunchtime conversations with Andrew Dancer, Iwao Hirose, Steffen Lauritzen, and Jim Oliver have all contributed in different ways to different parts of this book. Any mistakes are wholly of our own making.

Graham K. Taylor and Adrian L. R. Thomas
Oxford, November 2013

Funding

This book was completed during a period of sabbatical leave from Jesus College and the Department of Zoology, Oxford. The research leading to these results has received funding from the European Research Council under the European Community's Seventh Framework Programme (FP7/2007-2013)/ERC grant agreement no. 204513.

Contents

List of symbols

α aerodynamic angle of attack; intercept of a linear model

$\boldsymbol{\alpha}$ column vector with α_i as its ith entry

α_i intrinsic selective advantage of the ith allele

β slope of a linear model

$\hat{\beta}$ parameter estimate for β

$\boldsymbol{\beta}$ vector containing the parameters of a linear model

γ glide angle

$\boldsymbol{\delta}$ vector of residuals in a generalized least squares model

δ_i measurement error for the ith datum of X

ϵ_i measurement error for the ith datum of Y

$\boldsymbol{\epsilon}$ vector of residuals in an ordinary least squares model

η mass of blood as a proportion of body mass

θ angle of a pendulum from the vertical

Θ angular amplitude of a leg or pendulum

λ stride length; ratio of measurement error variances, $\sigma_\epsilon^2 / \sigma_\delta^2$

μ static coefficient of friction; fluid viscosity

ν branching factor of a hierarchical network

ρ fluid density

σ^2 error variance

$\sigma_\delta^2, \sigma_\epsilon^2$ measurement error variance for variables x, y

σ_q^2 equation error variance

υ resource consumed per unit volume of fluid

ϕ bank angle

Φ dimensionless function

ω angular velocity of leg at mid-stance

a scaling exponent; acceleration

$\mathbf{a}$	column vector with a_i as its ith entry
A_c	mean cross-sectional area of a capillary
a_i, a_j	total selective advantage of the ith, jth allele
A_i	mean cross-sectional area of a vessel of ith order
$\mathscr{R}$	aspect ratio of a wing
b	wingspan; scale factor in a scaling relationship
c	wing mean chord
C_D	aerodynamic drag coefficient
C_{Df}	friction drag coefficient
C_{Di}	induced drag coefficient
C_F	aerodynamic force coefficient
C_L	aerodynamic lift coefficient
d	distance from pivot to pendulum centre of mass
$\mathbf{D}$	diagonal matrix, $\mathbf{D} = \mathrm{diag}\,(\mathbf{p})$
D_i	induced drag
e	span efficiency factor
E	kinetic energy transferred in a downwash
$\dot{E}$	rate of kinetic energy transfer in a downwash
f	natural frequency of a mass-spring system
F	force; elliptic integral of the first kind
g	gravitational acceleration
$g_i(\mathbf{y}, \mathbf{x})$	inequality constraint, $g_i(\mathbf{y}, \mathbf{x}) \geq 0$
$h_j(\mathbf{y}, \mathbf{x})$	equality constraint, $h_j(\mathbf{y}, \mathbf{x}) = 0$
i	index of an allele, datum, network level, or constraint
I	moment of inertia of a pendulum
$\mathbf{I}$	identity matrix
j	index of an allele, datum, or constraint
k	integer constant; index of a genotype
K	leg stiffness
l	number of loci in a set; characteristic length
L	fundamental unit of length
L	aerodynamic lift
l_i	mean length of a vessel of ith order
m	body mass; pendulum mass; integer constant
$\dot{m}$	mass flow rate in downwash

M	fundamental unit of mass
n	ploidy; number of variables; sample size
N	square matrix with N_{ij} as its (i, j)th entry
N_c	number of capillaries
N_i	number of vessels of ith order
N_{ij}	frequency of association of ith allele with jth allele
p	momentum transferred to downwash; p-value
$\dot{p}$	rate of momentum transfer to downwash
p	row vector with p_i as its ith entry
P	aerodynamic power
P	lower triangular matrix satisfying $\mathbf{V} = \mathbf{PP}'$
p_i	relative frequency of the ith allele
Q	volume rate of flow
q_i	equation error for the ith datum of a dependent variable
r	turn radius
r^*	limiting turn radius
R	metabolic rate
R^2	R-squared value
Re	Reynolds number
s	total number of different alleles in a population
S	wing area
s_{XX}	sample variance of X
s_{XY}	sample covariance of X and Y
s_{YY}	sample variance of Y
t	time
T	fundamental unit of time
T	period of a pendulum; stride period
U	speed
U	matrix of explanatory variables in generalized least squares
U^*	best glide speed
u_c	mean flow speed in a capillary
U_s	sink rate
U_s^*	minimum sink rate
v	downwash speed
V	multilocus additive genetic variance in fitness; volume of fluid

V	unscaled error covariance matrix
v_i	vector used to form an inner product
w	relative fitness; exponent to be solved for
W	body weight
$w(\boldsymbol{y})$	relative fitness function
x	variable in a statistical model; exponent to be solved for
$\boldsymbol{x}$	vector of design variables
X	random variable
$\boldsymbol{X}$	matrix containing explanatory variables and a column of ones
x_i	true underlying value of the ith datum of a variable x
X_i	measured value of the ith datum of a variable x
y	variable in a statistical model; exponent to be solved for
$\boldsymbol{y}$	vector of performance objectives for natural selection
Y	random variable
$\boldsymbol{Y}$	vector of observations of a response variable
y_i	true underlying value of the ith datum of a variable y
Y_i	measured value of the ith datum of a variable y
z	exponent to be solved for
$\boldsymbol{Z}$	dependent variable in a generalized least squares model

1

Themes

1.1 Introduction

Half a century before Charles Darwin would propose his theory of natural selection, the Archdeacon of Carlisle, one William Paley, was using the appearance of design in nature to argue for the existence of a divine creator. The argument from design in Paley's *Natural Theology* (Paley, 1802) was not a new one, and his work might well have been consigned to obscurity following the Darwinian revolution, were it not for the eloquent analogy that he offered. Living organisms, Paley argued, functioned like clockwork—an analogy that has since been immortalized, if not as Paley had intended, in the title of Dawkins' book *The Blind Watchmaker* (Dawkins, 1986). In making his case, Paley offered a string of biomechanical examples, from the active stabilization of standing in humans, to the asymmetry of the wingbeat's upstroke and downstroke in birds (Paley, 1802). Paley's emphasis upon biomechanics should not surprise us, for the simple reason that there is no other realm of biology in which the link between form and function is so clear. What should surprise us—and for precisely the same reason—is how slight a role biomechanics has played in the subsequent development of evolutionary thought. Biomechanics offers an exceptionally fine lens through which to view pattern and process in evolution, and the purpose of this book is to hold that lens up to some fundamental questions in evolutionary biology.

One of the most important contributions that biomechanics can make to evolutionary biology is to elucidate the interaction between selection, phylogeny, and constraint. This interaction has proven surprisingly difficult to pin down. Even Gould, who thought extensively about such matters, could only offer a simple diagrammatic representation of the interaction (Gould, 2002), which he called the 'aptive triangle' (Figure 1.1). This is nothing more than a triangle labelled with the functional (i.e. adaptive), historical (i.e. phylogenetic), and structural (i.e. physical) 'causes' of a trait at its vertices. Different traits are imagined to sit at different points on the triangle according to the relative importance of each of the three causes. An important problem with this representation is that it treats logically distinct levels of explanation as if they were equivalent (cf. Tinbergen, 1963), and offers no insight into how—as 'causes'—they might actually interact. Fortunately, the physical and mathematical foundations of biomechanics allow us to go beyond this, and towards

Evolutionary Biomechanics. Graham Taylor & Adrian Thomas.
© Graham Taylor & Adrian Thomas 2014. Published 2014 by Oxford University Press.

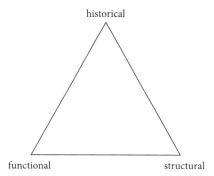

Figure 1.1 Gould's 'aptive triangle' (Gould, 2002), which depicts the 'causes' of traits as structural (i.e. physical), functional (i.e. adaptive), and historical (i.e. phylogenetic). A given trait is supposed to be located somewhere on the edges or interior of the triangle, according to the relative importance of the three causes. However, as these represent logically distinct levels of explanation, it is unclear how they are supposed to interact.

a formal framework for analysing the interaction between selection, phylogeny, and constraint. In this introductory chapter, we sketch out the framework that we will develop through the rest of this book.

1.2 Selection

Darwin's greatest contribution was to identify a natural process capable of producing the appearance of good design (Darwin, 1859): it is remarkable, therefore, that formal mathematical, rather than verbal, proof of the fact that natural selection has an optimizing tendency was still lacking almost a century and a half later (see Grafen, 2002; Batty et al., 2013). This is not a reflection upon the truth of the fact that natural selection has an optimizing tendency—for which there is a wealth of empirical support; not least from biomechanics. Rather it reflects the difficulties associated with formalizing Darwinism mathematically, which arise from the fact that selection acts ultimately at the level of competing alleles, whereas adaptation is defined at the organismal level. Many of the ongoing controversies over levels of selection stem from the conceptual difficulty in relating these two different levels of description.

The claim that natural selection has an optimizing tendency is logically distinct from the claim that a given trait of a given organism is in any sense optimal. To assert that natural selection has an optimizing tendency, as we have just done, is merely to ascribe directionality to that process. This underlying directional tendency was first proven mathematically by Fisher (1930, 1958), whose Fundamental Theorem of natural selection can be shown to describe one of two equal and opposite fluxes in the selective advantage of the alleles (see Chapter 2). These fluxes represent, respectively, the spread of those alleles that contribute positively to their own selective advantage, and the decline in the selective advantage of those same alleles as the adaptations that they confer become commonplace in the population. The net effect of these two fluxes is that the population becomes better adapted to its environment, even though the individuals that it contains are no better off than their ancestors in respect of their fitness relative to the population mean. Unfortunately, the difficulty of Fisher's presentation of the Fundamental Theorem—let alone of relating it to its consequences for adaptation—would consign his masterpiece to obscurity for decades. Worse than this, the Fundamental Theorem was misappropriated by Wright in support of his adaptive landscape metaphor (e.g. Wright, 1988), which expressed the flawed notion that natural selection has a tendency to climb hills whose height represents population mean fitness. Natural selection has no such tendency, and in fact population mean fitness will usually fluctuate around a growth rate of zero because of the effects of density-dependence.

In Chapter 2, we provide a new derivation and interpretation of Fisher's Fundamental Theorem in terms of this flux in the selective advantage of the alleles. We use our new interpretation of the Fundamental Theorem to redefine the adaptive landscape in a way that is consistent with the modern gene-centric view of evolution, but which also allows us to relate this to the appearance of organismal design. We achieve this by redefining the 'horizontal' axes of the landscape as performance objectives for natural selection. We define a performance objective as any quantity whose increase would be expected to enhance the selective advantage of an allele conferring that increase, in the hypothetical case that this increase could be effected without impacting performance in any other dimension. We take the 'vertical' axis of the landscape to represent the fitness of an individual, measured relative to the population mean, so that it can serve as a proxy for the expected rate of increase of any actual copy of an allele that the individual carries. As adaptations spread through the population, the fitness of individuals bearing those adaptations is brought closer to the population mean. Hence, since the height of the landscape is measured relative to population mean fitness, which we represent as 'sea level', the population does not actually climb the adaptive landscape. Rather, it remains standing near the waterline as the landscape itself slips beneath the waves. We call this the 'drowning landscape' model of adaptive evolution.

In practice, of course, it is not usually possible to improve performance in one dimension without degrading performance in another. Hence, since we will not usually be able to represent all of the possible performance objectives explicitly in the landscape, it follows that relative fitness will not actually increase monotonically with every performance objective. Instead, the time-varying shape of the landscape will reflect the trade-offs that exist between those performance objectives which are represented explicitly in the landscape and others which are not. Such trade-offs are the result of constraint, and it follows that the optimizing tendency of natural selection can only be understood with due regard to those constraints.

1.3 Constraint

Biomechanics is, in fact, particularly well suited to elucidating the interaction of selection and constraint. By way of illustration, we return to Paley's watchmaker analogy, but develop this a little further: '*Suppose I had found a watch upon the ground. ... we perceive ... that its several parts are framed and put together for a purpose, e.g. that they are so formed and adjusted as to produce motion, and that motion so regulated as to point out the hour of the day; that, if the several parts had been differently shaped from what they are, of a different size from what they are, or placed after any other manner, or in any other order, than that in which they are placed, either no motion at all would have been carried on in the machine, or none which would have answered the use, that is now served by it.*' (Paley, 1802). Suppose now that, instead of encountering a pocket watch, we had come across a collection of pendulum clocks of differing shapes and sizes. One could hardly fail to be impressed by the uniformity with which the clocks kept time, and we would be correct to attribute this on some level to human design. Upon unlocking their cases, however, it would become apparent that the pendulums of the taller clocks beat slower than those of the shorter ones, regardless of the details or workmanship of the rest of the mechanism. Plotting the period of each pendulum against its length, we would quickly see that all of the points fell along approximately the same line, scaling as the square root of pendulum length (Figure 1.2). The equation of that line is the result of the physical constraints that govern pendulum dynamics (see Chapter 3), although it is obvious that human design still plays an important part, even in this most highly constrained of examples. Clearly, we cannot understand the design of a pendulum clock by a human designer, nor the optimization of a living organism by natural selection, without understanding the underlying physical constraints.

To develop the argument a little further, the relationship between pendulum length and period defines the design space for clock pendulums, whilst the distribution of points within the design space is the result of human design. The majority of long-case clocks, for example, have a pendulum measuring almost exactly a metre in length. This is called a seconds pendulum, on account of the fact that an ideal pendulum measuring 0.994 m completes one swing every second under standard gravity. The seconds pendulum is a deliberate design feature, chosen for its aesthetic convenience. Its actual length is the result of historical contingency, however, because the division of the solar day into 86,400 seconds derives from the ancient Egyptians' use of twelve zodiacal signs to divide day and night, combined with the ancient Babylonians' penchant for working in base 60. In a world in which the ancients had preferred to work in base 10, so that every day was divided into 100,000 'seconds', the 'seconds pendulum' would measure 0.742 m instead of 0.994 m. It is therefore unlikely that we would find the same design preference for metre-long pendulums if we could rerun the tape of human history.[1]

---

[1] Or perhaps not, because the length of the seconds pendulum was originally used to define the metre (see Matthews, 2001). This definition was proposed in the late 17th century by Christiaan

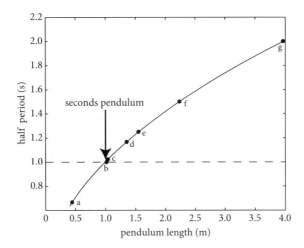

Figure 1.2 Half-period against pendulum length for seven historic pendulum clocks. The line shows the theoretical relationship for an ideal pendulum undergoing small amplitude oscillations under standard gravity. The constraint that is described by this line sets the available design space. The majority of long-case clocks are designed to have a pendulum that completes one swing every second: their position within the design space is indicated by the arrow and dashed line. a. Shuckburgh's equatorial instrument (John Arnold, 1793); b. Long-case clock with seconds pendulum (Crow of Faversham, *c.* 1790); c. Lantern clock (Nathaniel Hedge, *c.* 1740); d. Hanging clock (Sören Marcusen, 1799); e. Long-case clock (William Clement, *c.* 1680); f. Turret clock (William Clement, 1671); g. Observatory clock (Thomas Tompion, 1676). Data from Edwards (1977).

In an evolutionary context, we should like to be able to reach a similarly clear understanding of the interaction between physical constraint and natural selection. Consider, for example, the statistical relationship between peak ground reaction force and body mass that results if we measure these variables for a selection of animals hopping or trotting over a force plate (Figure 1.3a). This statistical relationship is constrained by the fact that the peak ground reaction force (F) cannot be less than gravitational acceleration (g) times body mass (m). This constraint can be expressed mathematically by the inequality $F \geq mg$. However, an even more general constraint is revealed if we also record the peak vertical acceleration (a) of our animals. What

---

Huygens and Sir Christopher Wren, and was subsequently adopted in 1790 by a committee established to recommend reform of the measurement system in revolutionary France. It was rejected a year later in favour of a standard defined as one ten-millionth of the quadrant of the arc of the Paris meridian—ostensibly because of an academic argument against mixing distance with time, but doubtless influenced by the several hundred thousand livres that the scientific establishment stood to receive to finance the necessary measurements of the Earth! By a remarkable coincidence, the two standards turned out to differ by just 0.3 mm. Ironically, today's standard also mixes distance with time, by defining the metre in relation to the speed of light and the vibration frequency of a caesium atom.

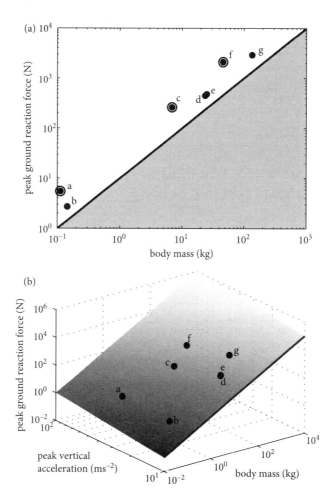

Figure 1.3 Measurements of the peak ground reaction forces exerted by seven species of mammal when trotting or hopping across a force plate. (a) Measured peak ground reaction force plotted against body mass. All of the points in (a) are constrained to lie above the line defined by the constraint that peak ground reaction force cannot be less than body weight. This inequality constraint is illustrated by the solid line and grey area on the graph. Circled points denote hopping gaits. (b) Measured peak ground reaction force plotted against body mass and (theoretical) peak vertical acceleration. All of the data points in (b) are constrained to lie upon the surface defined by the equality constraint imposed by Newton's second law. The solid line bounding one edge of the surface denotes the inequality constraint 'peak ground reaction force cannot be less than body weight'. a. *Dipodomys spectabilis*; b. *Rattus norvegicus*; c. *Macropus eugenii*; d. *Canis familiaris*; e. *Capra hircus*; f. *Macropus rufus*; g. *Equus caballus*. Data from Farley et al. (1993).

then emerges is a physical, rather than statistical, relationship between the variables, corresponding to Newton's second law of motion (Figure 1.3b). This physical constraint is defined by the mathematical equality $F = ma$, and there is obviously nothing that evolution or behaviour can do to escape it. Physical constraints such as these set the design space that evolution and behaviour are free to explore. In Chapter 3, we make use of these concepts in analysing the physical constraints upon three different terrestrial gaits, using the analytical approach known as dimensional analysis.

1.4 Scaling

Physical constraints are usually multidimensional, but it is commonplace in biology to summarize the statistical relationship between two physical variables x and y by fitting a bivariate scaling relationship of the form $y = bx^a$, where a and b are constants. In a biomechanical context, this inevitably means that some—perhaps most—of the scatter about the line of best fit will be due to the physical incompleteness of the fitted relationship. As a simple illustration, consider again the empirical relationship that exists between peak ground reaction force and body mass in animals hopping or trotting over a force plate. Species that are adapted for hopping exert consistently higher peak forces in relation to body mass than species that are adapted for trotting (Figure 1.3a). This systematic variation will be treated as random error if we fit a bivariate scaling relationship between peak ground reaction force and body mass. However, we know from Newton's second law that this error is actually attributable to unmeasured variation in peak vertical acceleration (Figure 1.3b). Statistical error of this sort is known as 'equation error', because it represents a deficiency in the fitted equation, rather than an error in the measured variables.

Equation error needs to be handled differently to measurement error, and this has some important consequences for the interpretation of scaling relationships in biology. Equation error is a ubiquitous feature of biological scaling relationships, but the errors-in-variables models that have been widely advocated for estimating scaling relationships (e.g. Rayner, 1985) fail to take account of it and are consequently inappropriate in this context. An important feature of equation error is that it is arbitrary how we choose to assign it between the variables. For example, should we treat the hopping animals in Figure 1.3a as having a large ground reaction force given their body mass, or a small body mass given their ground reaction force? This sounds like mere wordplay, but we will arrive at a different estimate of the statistical relationship between the variables according to which alternative we choose. Such ambiguity is appropriate, because there is no physical relationship that involves only the two variables that we are considering, and we should not therefore expect to find any unique statistical relationship between them. It follows that we should not read physical significance into the numerical value of their fitted scaling exponent. This deceptively simple point has some important ramifications.

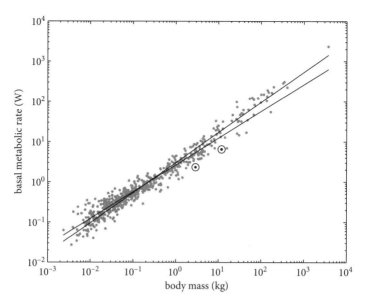

Figure 1.4 Measurements of basal metabolic rate (BMR) versus body mass for 626 species of mammal, from the dataset compiled by Savage et al. (2004b). The two lines show BMR scaling as either the two-thirds or three-quarters power of body mass. The difference between these lines pales into insignificance when compared to the variation in the data, because any relationship involving BMR and body mass is bound to involve a host of other variables besides. Attempts to distinguish whether 'the' slope of the statistical relationship between BMR and body mass is two-thirds or three-quarters miss this point entirely. The two encircled data points represent two species of egg-laying echidna (Tachyglossidae), which have an unusually low metabolic rate given their body mass. This presumably reflects systematic variation in certain of the other variables that govern metabolic rate, rather than random error.

For example, the long-running debate on metabolic scaling has all too often been concerned with asking whether metabolic rate scales with body mass raised to the power of two-thirds or three-quarters (Figure 1.4). However, because any relationship connecting metabolic rate and body mass is bound to involve a host of other variables, we should not behave as if there is some unique mapping from body mass to metabolic rate. Asking whether 'the' value of the scaling exponent of metabolic rate with body mass is two-thirds or three-quarters makes as much sense as asking whether the king of France is wise (see Strawson, 1950). The formulation of both questions presupposes the existence of something that does not exist: France has no monarch; neither is metabolic rate dictated by body mass. In Chapter 4, we therefore use metabolic scaling as a classic example of the statistical and epistemological problems that arise when we find ourselves summarizing a multivariable physical relationship by a bivariate scaling relationship.

1.5 Phylogeny

Equation error has an especially insidious effect in comparative studies, because the variation in important but unmeasured variables is likely to be correlated among related species. For example, echidnas (Tachyglossidae) have lower basal metabolic rates than other mammals of comparable body mass (Figure 1.4), which presumably reflects systematic variation in certain of the other variables that govern metabolic rate, rather than random error. Likewise, kangaroos (Macropodidae) exert higher peak ground reaction forces than other mammals of comparable body mass (Figure 1.3a), which presumably reflects shared variation in peak vertical acceleration associated with their hopping gait (Figure 1.3b). In fact, shared variation in important but omitted variables is expected to lead to error covariance among related species in any comparative study in which we attempt to fit a statistical relationship that does not involve all of the relevant variables. It is important to note that this error covariance is expected to disappear altogether if we fit a physically complete relationship: clearly, we do not expect a kangaroo to behave differently to any other animal in respect of Newton's second law. The apparent influence of phylogeny therefore depends upon how we model the world.

Failing to account for phylogenetic error covariance will not bias our estimate of the statistical relationship between variables, but it will increase the variance of the estimator. This has three important consequences. First, our estimate of the relationship will be noisier than it could be. Second, we will underestimate the width of the confidence intervals we fit. Third, we will incorrectly estimate the significance of our hypothesis tests. These are serious issues, so if we have reason to believe that we are fitting a statistical model with equation error in a study involving cross-species comparisons, then we will obviously need to take due account of error covariance. It is remarkable, therefore, that although statistical techniques for dealing with error covariance are well developed and widely used among evolutionary biologists, they have scarcely penetrated the biomechanics literature. For example, the extensive literature on comparative flight morphology has historically treated species as independent data points (but see Alerstam et al., 2007), despite the fact that there is a high degree of phylogenetic error covariance in the scaling relationships that have been studied. In Chapter 5, we use the scaling of wing area with body mass in birds as a canonical example to discuss phylogenetically controlled methods in the context of comparative biomechanics.

1.6 Form and function in flight

Comparative analyses controlling for phylogeny are a powerful tool for testing adaptation, but our ability to make theoretical predictions about the physical relationship between form and function will usually greatly exceed our ability to test these predictions empirically. For example, it is possible to make exquisitely detailed theoretical predictions about how wing loading (the ratio of weight to wing area) and aspect ratio (the ratio of wingspan squared to wing area) combine to determine average cross-country speed in a thermal-soaring sailplane or bird. Doing so requires a detailed model of thermal strength, size, and spacing, however (see e.g. Thomas, 1999). Furthermore, whereas we can ask sailplane designers what aspects of flight performance they have prioritized, and what constraints they have encountered in doing so, we will have no such luxury in evolutionary biomechanics. In practice, the only reliable information that we will often have available is a qualitative description of an organism's ecology or behaviour, such as 'species A uses soaring' or 'species B does not soar'. If qualitative statements such as these are our only available predictors of organismal function, then the most that we can hope to test in practice is a simple directional prediction about how natural selection is expected to shape organismal form. In arriving at these directional predictions, there is an obvious temptation to make whatever assumptions are necessary to use an existing piece of theory whose detailed predictions we then simplify. The problem with this approach is that we will lose generality if we inadvertently find ourselves having to make more restrictive assumptions than are necessary in order to arrive at the simple directional prediction that we will test. It is better to start from first principles, and to make our theoretical predictions as general as they possibly can be. In Chapter 6, we therefore combine simple physical reasoning with dimensional analysis to identify how flight performance is expected to vary with flight morphology. We then turn this around to make a series of directional predictions about how flight morphology is expected to change in response to selection for large transient forces, high glide speed, low sink rate, low power requirements, and high aerodynamic efficiency.

1.7 Adaptation in avian wing design

In Chapter 7, we undertake a formal phylogenetically controlled comparative analysis of whether and how flight morphology has responded to a variety of different ecological selection pressures in our comparative dataset of 450 species of bird. The first step is to identify a set of objectively identifiable behavioural or ecological characters that we expect will be associated with selection for the different aspects of flight performance that we already identified in Chapter 6. We do this by identifying behavioural or ecological characters that are sufficiently obvious that they can be scored on the basis of the detailed species accounts that appear in the regional handbooks for birds—characters such as soaring over land, sally hunting, or submerging fully to feed. We score these characters in a binary fashion, simply noting whether or not they are recorded for the species in question, and making no subjective judgement as to the relative importance of that character to that species. We then use these binary ecological characters as predictors in a phylogenetically controlled analysis of the variation that exists in different aspects of wing morphology, such as wing area and wingspan, controlling for the scaling associated with variation in body mass. The results of this analysis are consistent with our directional predictions about how flight morphology should respond to selection for different aspects of flight performance.

However, they are surprising to the extent that several key ecological predictors, such as the use of migration, are not statistically significant in any of the analyses. This might simply be because our comparative approach is a relatively blunt instrument for detecting the effects of selection; especially given our binary scoring of the ecological predictors. However, it is also possible that the absence of some of the effects that we expected to see is due to the trade-offs that exist between different aspects of flight performance. Such trade-offs are the subject of Chapter 8.

1.8 Trade-offs: selection, phylogeny, and constraint

In our 'drowning landscape' model of adaptive evolution, the 'horizontal' axes of the landscape are defined as performance objectives for natural selection. Conceptually, these performance objectives represent any quantity whose increase would be expected to enhance the selective advantage of an allele conferring that increase, were it possible to effect such an increase without impacting performance in any other dimension. Hence, although the optimizing tendency of natural selection is usually discussed with respect to fitness maximization, which is a constrained single-objective optimization, it may also be thought of as a constrained multi-objective optimization with respect to the various conflicting performance objectives. Since we will not know in general how natural selection weights these performance objectives in a given species, it is most natural to analyse the outcome of this multi-objective optimization using the concepts of dominated and non-dominated solutions, and Pareto optimality.

A phenotype may be said to dominate another phenotype with respect to a given set of performance objectives, if and only if it performs as well or better than the other phenotype on every single performance objective. The subset of phenotypes that is not dominated by any other is called the Pareto set, and each of the phenotypes that this set contains may be said to be Pareto optimal with respect to the specified set of performance objectives. Optimality in the Pareto sense is therefore quite different in meaning to the usual sense of the word optimality. However, it is always possible to hypothesize different sets of weights for the various conflicting performance objectives that would make any member of the Pareto set optimal in the usual sense of the word. As an example of the kind of quantitative analysis that can be achieved using these concepts, we compare the flight morphology of soaring petrels and albatrosses (Procellariiformes) with the flight morphology of broad-winged raptors (Accipitriformes). We identify three key performance objectives for soaring flight, and analyse the detailed trade-offs that exist between these given the known theoretical constraints upon flight mechanics. We then use our morphological dataset for birds to predict how each species of soaring bird performs with respect to the three soaring flight performance objectives.

Procellariiformes and Accipitriformes occupy different regions of the available morphospace, which map on to different regions of the performance space defined by the three soaring performance objectives. By identifying the Pareto sets with respect to different combinations of soaring objectives, we are able to show that Procellarii-formes in general and albatrosses (Diomedeidae) in particular are typically Pareto optimal with respect to the performance objectives that are important for straight soaring flight. In contrast, Accipitriformes in general, and vultures (Cathartidae, Gypaetinae, Aegyptiinae) in particular, are typically Pareto optimal with respect to performance objectives that are important for turning flight performance. The great majority of species in both groups are Pareto optimal when all three soaring performance objectives are considered, which indicates that most of the variation in flight morphology can be explained in principle by variation in how these three conflicting performance objectives are weighted in the different species.

1.9 Conclusion

In summary, the diversification of species can be thought of as the process of finding alternative trade-offs among conflicting performance objectives in organisms subject to similar constraints. True evolutionary innovation is brought about when a change in either the genes or the environment introduces new performance objectives for natural selection. Thus, an analysis of multi-objective optimization pulls together the three key threads of this book: selection, phylogeny, and constraint. Evolutionary biomechanics is the study of the interaction of these three key influences upon adaptive evolution.

2

Selection

2.1 Introduction

It is a truth almost universally acknowledged, that individual organisms act as if try-ing to maximize their fitness.[1] This, Darwin argued, would be the inevitable outcome of natural selection, and his fitness maximization principle has been invoked ever since to explain the appearance of organismal design in disciplines ranging from behavioural ecology to biomechanics. Ideas about optimization lie at the heart of adaptationism, and are of particular importance in evolutionary biomechanics, with its focus upon relating form to function in the context of physical constraint. It is therefore surprising to find that the very idea that natural selection has an optimiz-ing tendency, which Grafen has called the 'individual-as-maximizing-agent analogy', has until recently lacked formal theoretical justification in the literature (see Grafen, 2002, 2007, 2008; Batty et al., 2013). Maximization principles do not emerge easily from the mathematics of population genetics, and those which have been found turn out to be too restrictive in their assumptions (e.g. Kimura, 1958; Kingman, 1961), or to maximize a quantity that is too esoteric to supply the general principle that we require (e.g. Iwasa, 1988; Ewens, 1992).

As a result, we find ourselves in the unexpected position of beginning this book with an analysis of the theoretical basis for assuming an optimizing tendency in evol-ution. In so doing, we aim to unite the rest of this book, with its emphasis upon whole-organism biology, with the underpinnings of population genetics. This is not a mere tidying-up exercise, because the nature of the underlying optimization pro-cess must obviously affect the kinds of solution that we can expect to emerge at the organismal level. There are good, bad and indifferent schemes for optimization in engineering design, as measured by their speed, efficiency and effectiveness, but the question of where on this spectrum natural selection falls has scarcely been asked—let alone answered. We will not reach a definite answer here, but will begin to erect the framework that is needed to tackle such questions.

---

[1] With apologies to Jane Austen, whose opening remarks to *Pride and Prejudice* concerned sexual selection and the descent of man.

Grafen's 'formal Darwinism project' (summarized in Grafen, 2008; Batty et al., 2013) aims to provide a formal mathematical justification for adaptationism, by establishing links between two quite different mathematical abstractions: the problem statement of an optimization program, and the covariance selection mathematics of the Price equation (Grafen, 2002). The abstract nature of these links may not yield immediate intuition into the underlying optimization process, but the essential features of that process are captured by Fisher's Fundamental Theorem of natural selection (Fisher, 1930, 1958). The Fundamental Theorem is not itself a fitness maximization principle, but when properly understood—which we will argue it has not been—it supplies a law of increase that can indeed be used to justify the 'individual-as-maximizing-agent' analogy.

The Fundamental Theorem is conventionally understood as defining the 'partial' increase in population mean fitness that is expected to result from changes in gene frequency under natural selection (Price, 1972; Ewens, 1989; Edwards, 1994; Frank, 1997; Lessard, 1997; Grafen, 2003; Ewens, 2004; Okasha, 2008; but see Ewens, 2011). This is a valid interpretation, but its incompleteness is troubling. Moreover, an increase in population mean fitness, partial or otherwise, is not what organismal biologists have in mind when they talk of optimization (Grafen, 2008), and a description of natural selection based upon population mean fitness does not sit comfortably with the modern gene-centric view of evolution. We therefore begin this chapter by providing a new derivation of the Fundamental Theorem, which clarifies its meaning in the context of gene-level selection. Our derivation shows that the Fundamental Theorem is correctly understood as predicting one of two equal and opposite evolutionary fluxes, and thereby characterizes exactly and completely the rate at which selective turnover occurs in a population.[2] This conclusion is consistent with the results of recent 'fitness flux' models of evolution, derived from the asymptotics of diffusion equations used in statistical physics (Mustonen and Lässig, 2009; Zhang et al., 2012). This emphasis upon flux is important, because it allows us to redefine Wright's flawed but highly influential metaphor of the adaptive landscape (Wright, 1932) in a way that is consistent with the results of the Fundamental Theorem.

In our revised model of the adaptive landscape, the 'vertical' ordinate represents individual fitness relative to the population mean. Because this quantity is measured relative to the population mean, it has the important property that it can serve as a proxy for the expected rate of increase in frequency of any actual copy of an allele that the individual carries, in contrast to an absolute measure of fitness, such as number of successful gametes. Each of the 'horizontal' axes represents a quantitative phenotypic property of the individual that can be regarded as a performance objective for natural selection. We define a performance objective as any quantity whose increase would be expected to enhance the selective advantage of an allele conferring that increase, in the hypothetical case that this increase could be effected without impacting performance in any other dimension. If all conceivable performance objectives were used as

---

[2] A flux can be thought of as a bulk rate of change; in this case, summed over all the alleles.

horizontal axes, then the time-varying shape of the landscape would simply depend upon how the different performance objectives combined to determine relative fitness. The regions of the landscape that it is actually possible to occupy would then be determined separately by the constraints upon the problem. In practice, we will only ever be able to work with a lower-dimensional representation of the landscape. In this case, the shape of the landscape will depend also upon the trade-offs between those performance objectives which are represented explicitly in the landscape and those which are not. These trade-offs are implicit in the constraints upon the problem, and can be identified from first principles in problems of a biomechanical nature.

Because individual fitness is measured relative to the population mean, it will prove useful to picture population mean fitness as the level of the sea, relative to which the height of the landscape is measured. This representation allows us to link the underlying flux in the selective advantage of the alleles to the apparent maximization of fitness by individuals. At any given time, the inherent directionality of this flux is expected to cause an increasing proportion of the population to sit within the higher occupied regions of the landscape. This will tend to bring the fitness of those individuals occupying these high regions of the landscape closer to the population mean. Hence, because the height of the landscape above sea level represents individual fitness relative to the population mean, it follows that the landscape itself must slip beneath the waves, until the highest occupied peaks have been brought down to the sea. Those individuals that cling on at the waterline are no better off than their ancestors insofar as their relative fitness is concerned. They are, however, likely to be about as well adapted as they can be to the prevailing conditions, given the scope of the available genetic variation. Hence, given sufficient elapsed time, sufficient genetic variation, and sufficient environmental stability, individuals will indeed appear to act as if maximizing their fitness.

2.2 Fisher's Fundamental Theorem of natural selection

Fisher stated the Fundamental Theorem verbally as follows: '*The rate of increase in fitness of any organism at any time is equal to its genetic variance in fitness at that time.*' (Fisher 1930; the italics are his own). The abstruseness of that statement, the difficulty of its derivation, and the consequent obscurity of what Fisher meant by '*the rate of increase in fitness*' have caused his theorem to be misunderstood and misinterpreted to the present day. Nevertheless, if anything is clear from Fisher's verbal statement of his theorem, it is that he understood it to describe a general ('*of any organism*'), instantaneous ('*at any time*'), and exact ('*is equal to*') constraint upon the dynamics of natural selection. Ultimately, this constraint arises because Fisher treated evolution as referring specifically to changes in the relative frequency of the alleles, and used these changes as a measure of the selective advantage of the allele. Changes in allele frequency must always sum to zero, from which it follows that evolutionary change represents the dynamics of a zero-sum situation.

As we now show, the term corresponding to Fisher's '*rate of increase in fitness*' in his own derivations (Fisher, 1930, 1958) is one of two evolutionary fluxes that arise naturally when one of the fundamental mathematical constraints upon the problem is differentiated with respect to time. The first of these two fluxes represents the spread of selective advantage through the population under natural selection, as a direct result of changes in allele frequency; the second of these fluxes represents the inevitable diminishment of the selective advantage of the alleles as they spread through the population. This second flux reflects the fact that the evolutionary change occurring under natural selection gets the individuals comprising the population nowhere, on average, in terms of their fitness relative to the population mean. Once this point is understood, it becomes possible to see why the Fundamental Theorem captures exactly and completely the expected rate of selective turnover in a population. We begin by providing a simple derivation of Fisher's Fundamental Theorem, which helps to uncover several interesting aspects of the theorem that do not appear to have been appreciated previously.

2.2.1 Derivation of the Fundamental Theorem

We consider a population of individuals whose properties are affected by the information contained in their genotypes. Each genotype comprises an integer number of discrete strings of information, called alleles. Each allele is stored at a discrete location, called its locus, which is capable of storing only one allele at a time from among the subset of alleles that it can possibly store. We assume that the copies of the alleles that make up a single genotype are stored by n sets of l loci, where n is called the ploidy of the population. Identical strings of information occurring at different loci are treated as different alleles, and we assume that a total of s different alleles occur over all loci in the population as a whole. The relative frequency of the ith allele (p_i) is defined as its frequency in the population, expressed as a proportion of all of the alleles occurring at the same locus[3]. Because these proportions must sum to one at every locus, it follows that their sum over all loci is equal to the number of loci, such that $\sum_{i=1}^{s} p_i = l$. We assume that the relative frequencies of the alleles vary only through the births and deaths of the individuals that carry them. In other words, we are modelling only the spread of existing alleles, and not the appearance of new alleles through mutation. We further assume that the alleles that an individual carries are faithful copies of alleles carried by its parent(s), and assume that every allele has an equal chance of being copied from parent to offspring. Finally, we assume that the

---

[3] The relative frequencies of the alleles may be defined with or without weighting by the mean reproductive value of their bearers (see Price, 1972). If allele frequencies are not weighted by reproductive value, then their growth rates in the population will depend upon such factors as the age and sex of their bearers. Such effects will be eliminated if the allele frequencies are weighted by the mean reproductive value of their bearers. Clearly, this distinction affects the interpretation of selective advantage, as defined by Eq. 2.1 below. It does not, however, affect the mathematics of the derivation that follows, which holds in either case.

population is very large and that alleles do not disappear suddenly from the population, such that their relative frequencies can be treated as differentiable functions of time (t).

An increase in the proportion of one allele must always be balanced by a decrease in the proportion of one or more of the other alleles occurring at the same locus. Mathematically, this means that the constraint $\sum dp_i/dt = 0$ must always be satisfied, whether the summation is taken over all of the alleles occurring at a single locus, or over all of the alleles occurring at all loci. It follows, in the latter case, that $\sum_{i=1}^{s} dp_i/dt = 0$. This constraint arises naturally whenever evolution is treated as referring to changes in the relative frequency of the alleles, and holds irrespective of changes in population size by virtue of the definition of a proportion. We note in passing that this result can be obtained directly by differentiating the constraint $\sum_{i=1}^{s} p_i = l$ with respect to time, given that the number of loci (l) is a constant. We will use the same technique again later to establish a deeper constraint upon the evolutionary dynamics. This deeper constraint is essential to understanding the meaning of the Fundamental Theorem.

We begin by defining the total selective advantage (a_i) of the ith allele as

$$a_i = \frac{1}{p_i}\frac{dp_i}{dt}, \tag{2.1}$$

which is simply the instantaneous rate of increase, sometimes called the Malthusian growth parameter, of the relative frequency of the ith allele in the population. In general, a_i is a time-varying quantity, which may be positive or negative depending upon whether the allele is increasing or decreasing in the population. Fisher referred to a_i as the 'average excess' of the ith allele in respect of fitness (Fisher, 1958), but average excess can be defined in respect of any quantitative trait, and its more general definition obscures the particular meaning of the quantity in the context of the Fundamental Theorem. In fact, the exposition of the Fundamental Theorem is greatly clarified if we understand a_i simply to represent the total selective advantage of the ith allele, and do not trouble ourselves with Fisher's general definition of average excess as a weighted mean. Eq. 2.1 can be easily rearranged to show that $p_i a_i = dp_i/dt$, and because $\sum_{i=1}^{s} dp_i/dt = 0$, it follows immediately that the constraint $\sum_{i=1}^{s} p_i a_i = 0$ must always hold. The same constraint holds if the summation is only taken over the alleles occurring at a single locus.

No allele occurs in isolation in a genotype: on the contrary, every allele is found in combination with ($nl - 1$) others. It follows that the total selective advantage (a_i) of the ith allele must be due not only to the properties that it confers, but also to the properties conferred by any alleles with which it is statistically associated in the population. Alleles need not combine linearly in their genotypic effects, but because a_i already represents an average over all genotypes, it must be possible to relate it to the average effects of the alleles in the genotypes occurring in the population. In other words, we must be able to partition the total selective advantage of the ith allele into an additive contribution due to the allele itself, and an additive contribution due to the other alleles with which it is statistically associated in the population (including any other copy of the ith allele in organisms with ploidy $n \geq 2$). It is important to emphasize that we are not thereby ignoring the effects of the environment, because these are implicit in the effects of the alleles, which the environment modulates.

Box 2.1 Worked example of the frequencies of association of the alleles.

As an aid to understanding how the frequencies of association (N_{ij}) of the alleles are calculated, we provide here a simple worked example that nevertheless captures the complexities of the problem. This is done for a diploid organism $(n = 2)$ with two loci $(l = 2)$ and two different alleles occurring at each locus. It follows that there are only four different alleles in the population $(s = 4)$, although we may expect that $s \gg nl$ in any real population. Hypothetical values of the relative frequencies of the genotypes corresponding to the nine possible combinations of the four alleles (A_1, A_2, A_3, A_4) are given below, for a population with non-random mating:

locus 1	locus 2	relative frequency
A_1A_1	A_3A_3	$P_{1133} = 0.06790$
A_1A_2	A_3A_3	$P_{1233} = 0.18600$
A_2A_2	A_3A_3	$P_{2233} = 0.04480$
A_1A_1	A_3A_4	$P_{1134} = 0.22800$
A_1A_2	A_3A_4	$P_{1234} = 0.24400$
A_2A_2	A_3A_4	$P_{2234} = 0.16000$
A_1A_1	A_4A_4	$P_{1144} = 0.00300$
A_1A_2	A_4A_4	$P_{1244} = 0.00090$
A_2A_2	A_4A_4	$P_{2244} = 0.06460$

The frequency of association of allele A_1 with each of the alleles is:

$$N_{11} = (P_{1133} + P_{1134} + P_{1144})/p_1$$
$$N_{12} = (P_{1233} + P_{1234} + P_{1244})/2p_1$$
$$N_{13} = (2P_{1133} + P_{1134} + P_{1233} + (1/2)P_{1234})/p_1$$
$$N_{14} = (2P_{1144} + P_{1134} + P_{1244} + (1/2)P_{1234})/p_1$$

where p_1 is the relative frequency of allele A_1 in the population. The other N_{ij} can be obtained similarly. Hence, defining **p** as the row vector with p_i as its ith entry, and **N** as the square matrix with N_{ij} as its (i, j)th entry, we have:

$$\mathbf{N} = \begin{bmatrix} 0.48593 & 0.51407 & 1.09289 & 0.90711 \\ 0.82177 & 0.17823 & 0.92830 & 1.07170 \\ 1.30603 & 0.69397 & 0.58114 & 0.41886 \\ 1.15004 & 0.84996 & 0.44437 & 0.55563 \end{bmatrix}$$

$$\mathbf{p} = [\,0.61470 \quad 0.38454 \quad 0.51438 \quad 0.48485\,]$$

for the genotype frequencies given above. It is easy to verify numerically using this example that $\sum_{i=1}^{s} p_i N_{ij} = p_j (nl - 1)$. This is an important result, which we prove in general in Box 2.3, where we use it to show that the average intrinsic selective advantage of the alleles must always be zero over all loci.

It is reasonable to call that component of a_i which is attributable to the ith allele the intrinsic selective advantage of the ith allele (α_i). Furthermore, the additive contribution of the jth allele to the total selective advantage of the ith allele can be nothing other than its own intrinsic selective advantage (α_j) multiplied by the frequency with which it is associated with the ith allele in the population (N_{ij}). This last quantity needs careful definition, because we are only interested in calculating the frequency of association in genotypes that we already know to contain the ith allele. An accurate definition can be given by considering a sampling experiment in which we select one set of loci at random from each individual in the population, and retain any genotype in which the chosen set of loci contains a copy of the ith allele. Considering only those genotypes retained in the resulting sample, the frequency N_{ij} is then the expected number of copies of the jth allele accompanying the identified copy of the ith allele in a genotype selected at random from such a sample. It is, of course, possible to define N_{ij} arithmetically without recourse to the device of a sampling experiment: Box 2.1 provides a simple worked example of the calculation for readers who are interested in understanding how the N_{ij} relates to the relative frequencies of the different genotypes in the population.

With this definition of the N_{ij}, we may immediately write down an equation of the form

$$a_i = \alpha_i + \sum_{j=1}^{s} N_{ij}\alpha_j \qquad (2.2)$$

for every one of the alleles occurring in the population, where in general the α_i and N_{ij} are time-varying quantities. Notational differences obscure the comparison, but the resulting set of simultaneous equations—which we have written down from first principles—are exactly equivalent to the simultaneous equations that are used in the population genetics literature to define what is called the 'multilocus average effect' of an allele upon fitness (e.g. Ewens, 2004).[4] This means that what we have called the intrinsic selective advantage of the ith allele is exactly equivalent to the multilocus average effect of the ith allele in respect of fitness. The advantage of treating a_i as the total selective advantage of the ith allele, and α_i as its intrinsic selective advantage is that this clarifies the nature of the relationship between these quantities in a way that the more general definitions of average excess and average effect do not. Crucially, it avoids the need to define what is meant by 'fitness', which may be a property either of an individual or a genotype, by focussing instead upon selective advantage, which is a property of an allele.

---

[4] For example, Eq. 2.2 is the continuous-time equivalent of Eq. 7.43 in Ewens (2004). In our notation, the righthand side of Eq. 7.43 of Ewens (2004) is equivalent to $p_i a_i$ (see Eq. 12 in Lessard, 1997, for confirmation of this), while the left-hand side is equivalent to p_i times the right-hand side of our Eq. 2.2.

Given the assumptions made at the outset, there are only a few ways in which an allele can help to confer upon itself a positive selective advantage. First, an allele can confer properties upon its bearer that directly promote the bearer's own survival and reproduction. This includes the effects of natural selection and sexual selection as they were understood by Darwin. Second, the properties that an allele confers upon its bearer can have differential effects upon the survival or reproduction of another individual that depend upon the expected frequency of the allele in that individual. This encompasses the effects of kin selection and green beard genes. Third, the properties that an allele confers upon its bearer can promote favourable statistical associations with other alleles. This is the basis, for example, of Fisher's runaway process of sexual selection. Thus, far from needing modification to take account of the effects of kin selection, or indeed of sexual selection, the Fundamental Theorem already includes them. The fact that the Fundamental Theorem takes full account of the effects of kin selection has recently been proven mathematically by Bijma (2009). It is interesting to note that the same result can be intuited immediately if α_i is treated as the intrinsic selective advantage of an allele, and not as the average effect of the allele upon fitness, which is an estimate of the same quantity.

The simultaneous equations corresponding to Eq. 2.2 are not linearly independent, and in order to supply a unique solution for the α_i, it is usual to impose the constraint that the products $p_i\alpha_i$ must sum to zero for all of the alleles occurring at any given locus (see e.g. Ewens, 2004; Lessard, 1997; Ewens, 2011). This convention makes good sense in light of the fact that the products $p_i a_i$ must sum to zero for all of the alleles occurring at any given locus, but it is perfectly possible mathematically (see also Ewens, 2011) to impose different constraints upon the sum of the products $p_i\alpha_i$ at different loci. Nevertheless it can be shown that Eq. 2.2 implies that the constraint $\sum_{i=1}^{s} p_i\alpha_i = 0$ always holds, where the summation is taken over all loci (see Box 2.3). As we show in the next section, this constraint is the key to unlocking the meaning of the Fundamental Theorem.

It is clear by inspection of Eq. 2.2 that the total selective advantage (a_i) of the ith allele is a linearly increasing function of the intrinsic selective advantage (α_i) of the ith allele. This does not guarantee that a_i and α_i will have the same sign, because the value of a_i also depends upon the effects of other alleles with which the ith allele is associated in the population (Eq. 2.2). However, the summation $\sum_{i=1}^{s} p_i a_i \alpha_i$ is equal simply to the covariance between the total and intrinsic selective advantage of the alleles. Substituting the right hand side of Eq. 2.1 into this summation, we have

$$\sum_{i=1}^{s} p_i a_i \alpha_i = \sum_{i=1}^{s} \frac{dp_i}{dt} \alpha_i, \tag{2.3}$$

which in words says that the average flux in the intrinsic selective advantage of the alleles due to natural selection (i.e. the right-hand side) is equal to the covariance of their total and intrinsic selective advantage (i.e. the left-hand side). Aficionados will recognize this result as being closely related to the Price equation (Price, 1970), from which Eq. 2.3 can also be derived (see also Frank, 1997).

Box 2.2 Proof of the directionality of natural selection.

The summation $\sum_{i=1}^{s} p_i a_i \alpha_i$ can be rewritten in matrix form as

$$\sum_{i=1}^{s} p_i a_i \alpha_i = \boldsymbol{\alpha}^{\mathrm{T}} \mathbf{D} \mathbf{a}$$

where $\boldsymbol{\alpha}$ is the column vector with α_i as its ith entry, $\mathbf{a}$ is the column vector with a_i as its ith entry, and $\mathbf{D} = \mathrm{diag}\,(p_1, \ldots, p_s)$ is the diagonal matrix with p_i as the ith entry on its leading diagonal. The superscript T denotes the matrix transpose. Eq. 2.2 can be rewritten in matrix form as

$$\mathbf{a} = (\mathbf{I} + \mathbf{N})\,\boldsymbol{\alpha}$$

where $\mathbf{I}$ is the identity matrix and $\mathbf{N}$ is the square matrix with N_{ij} as its (i,j)th entry. We note that when premultiplied by $\mathbf{D}$, this equation becomes the continuous-time equivalent of Eq. 7.44 in Ewens (2004). Combining the two previous equations, we have:

$$\sum_{i=1}^{s} p_i a_i \alpha_i = \boldsymbol{\alpha}^{\mathrm{T}} \mathbf{D}\,(\mathbf{I} + \mathbf{N})\,\boldsymbol{\alpha}.$$

The matrix $\mathbf{D}\,(\mathbf{I} + \mathbf{N})$ can be defined by reconsidering the sampling experiment used to define the $N_{i,j}$ in the main text, wherein we select one set of loci at random from each individual in the population, and retain any genotype in which the chosen set of loci contains a copy of the ith allele. The (i,j)th entry of the matrix $\mathbf{D}\,(\mathbf{I} + \mathbf{N})$ is then the probability that the chosen set of loci contains a copy of the ith allele (i.e. the relative frequency, p_i, of the ith allele), multiplied by the total number of copies of the jth allele that are expected to be present in a genotype selected at random from such a sample.

As we now show, it follows from the above that the matrix $\mathbf{D}\,(\mathbf{I} + \mathbf{N})$ is positive semi-definite. This means that $\boldsymbol{\alpha}^{\mathrm{T}} \mathbf{D}\,(\mathbf{I} + \mathbf{N})\,\boldsymbol{\alpha} \geq 0$ for any vector $\boldsymbol{\alpha}$ of real numbers (e.g. Horn and Johnson, 2013), and hence that the summation $\sum_{i=1}^{s} p_i a_i \alpha_i \geq 0$, which was Fisher's key result. This is because the matrix $\mathbf{D}\,(\mathbf{I} + \mathbf{N})$ is a Gram matrix of inner products of a set of vectors $\{v_1, \ldots, v_s\}$, and every Gram matrix is positive semidefinite (Horn and Johnson, 2013). The relevant vectors $\{v_1, \ldots, v_s\}$ are of length equal to the number of different genotypes occurring in the population. The kth entry of the ith vector (v_i) is the square root of the product of the number of copies of the ith allele occurring in the kth genotype and the expected frequency of that genotype in the sampling experiment described above for the ith allele. Interested readers may want to verify this fact for the worked example in Box 2.1.

It is shown mathematically in Box 2.2 that the covariance $\sum_{i=1}^{s} p_i a_i \alpha_i$ is always non-negative. Intuitively, this is because alleles at other loci must be as likely on average to be positive as negative in their contribution to the total selective advantage of the ith allele. We may therefore use Eq. 2.3 to write

$$\sum_{i=1}^{s} \frac{dp_i}{dt} \alpha_i \geq 0, \tag{2.4}$$

which in words says that the average flux in the intrinsic selective advantage of the alleles due to natural selection is always non-negative. We discuss the meaning of this flux in the next section, but note that this inequality is the source of the directionality that lies at the heart of the Fundamental Theorem—and indeed of natural selection itself. It is, in other words, the very basis upon which a formal theory of adaptationism may be built (see Grafen, 2002; Batty et al., 2013). Fisher (1958) proved this directionality by relating his '*rate of increase in fitness*' to the additive genetic variance in fitness, but the explanation that is given in Box 2.2 shows that it is possible to derive the same result without ever needing to define fitness.

In order to relate our results more directly to Fisher's, we need only note that the quantity $n \sum_{i=1}^{s} p_i a_i \alpha_i$ is equal to the multilocus additive genetic variance in fitness (V), which is a standard quantity in population genetics (see e.g. Ewens, 2004). Substituting the identity $V = n \sum_{i=1}^{s} p_i a_i \alpha_i$, it is immediate from Eq. 2.3 that

$$V = n \sum_{i=1}^{s} \frac{dp_i}{dt} \alpha_i \geq 0 \qquad (2.5)$$

where the term on the right-hand side of the equality is what Fisher called '*the rate of increase in fitness*' (Fisher, 1930, 1958) and is equal to the additive genetic variance in fitness, which is always non-negative by virtue of being a variance. This completes the derivation of Fisher's Fundamental Theorem of natural selection.

2.2.2 Meaning of the Fundamental Theorem

Following Price (1972), Fisher's '*rate of increase in fitness*' is conventionally regarded as the 'partial' change in population mean fitness that would result from changes in allele frequency if the effects of the alleles upon individual fitness were held constant (Ewens, 1989; Edwards, 1994; Frank, 1997; Lessard, 1997; Grafen, 2003; Okasha, 2008; but see Ewens, 2011). This interpretation is tenable if the α_i in Eq. 2.5 are treated as the multilocus average effects of the alleles upon individual fitness. Its usefulness is questionable, however, because the average effects of the alleles are clearly not constant—depending as they do upon the genetic composition of the population and the effects of the environment (Ewens, 1989). This obvious deficiency has led to the general sense that the Fundamental Theorem says less about natural selection than its name suggests (but see Grafen, 2003; Okasha, 2008). Furthermore, it is mathematically idiosyncratic, if not incorrect, to use the Fundamental Theorem to draw conclusions about 'partial' changes in the rate of change in population mean fitness under the assumption that the average effects of the alleles are held constant (Ewens, 1989; Edwards, 1994; Ewens, 2011). As we now argue, however, this is not really what the Fundamental Theorem is about.

Box 2.3 Proof of the key constraint of the Fundamental Theorem.

We aim to prove that $\sum_{i=1}^{s} p_i \alpha_i = 0$, subject only to Eqs. 2.1 and 2.2 and the definitions of the terms therein. Multiplying both sides of Eq. 2.2 by p_i and summing over all loci yields

$$\sum_{i=1}^{s} p_i a_i = \sum_{i=1}^{s} p_i \alpha_i + \sum_{i=1}^{s} \sum_{j=1}^{s} p_i N_{ij} \alpha_j.$$

As shown in the main text, Eq. 2.1 implies that $\sum_{i=1}^{s} p_i a_i = 0$. We may therefore rewrite the preceding equation as

$$\sum_{i=1}^{s} p_i \alpha_i = -\sum_{i=1}^{s} \sum_{j=1}^{s} p_i N_{ij} \alpha_j$$

and switch the order of the summations to yield

$$\sum_{i=1}^{s} p_i \alpha_i = -\sum_{j=1}^{s} \alpha_j \sum_{i=1}^{s} p_i N_{ij}$$

where N_{ij} denotes the frequency with which the jth allele is associated with the ith allele in the population. It follows that the summation $\sum_i p_i N_{ij}$ taken over all of the alleles occurring at a single locus must equal the frequency of the jth allele multiplied by the number of alleles accompanying the ith allele at that locus. Where the summation is taken over the locus of the jth allele, the number of alleles accompanying the ith allele is one less than the ploidy $(n-1)$. At the other $(l-1)$ loci, the number of alleles accompanying the ith allele is equal to the ploidy (n). Summing over all loci, we therefore have

$$\sum_{i=1}^{s} p_i N_{ij} = \left[(n-1) + (l-1)\, n \right] p_j$$

$$= (nl - 1)\, p_j.$$

Substituting this result into the equation above, we therefore have

$$\sum_{i=1}^{s} p_i \alpha_i = -(nl-1) \sum_{j=1}^{s} p_j \alpha_j \equiv -(nl-1) \sum_{i=1}^{s} p_i \alpha_i.$$

This equation only holds if $\sum_{i=1}^{s} p_i \alpha_i = 0$, which completes the proof. This result can be stated more elegantly by writing the summations in matrix form, with $\mathbf{p}$ as the row vector with p_i as its ith entry, $\boldsymbol{\alpha}$ as the column vector with α_i as its ith entry, and $\mathbf{N}$ as the square matrix with N_{ij} as its (i,j)th entry. By definition, the summation $\sum_{i=1}^{s} p_i N_{ij} = (nl-1)\, p_j$ is just the jth entry of the row vector $\mathbf{pN}$. We therefore have $\mathbf{pN} = (nl-1)\, \mathbf{p}$, which implies that $\mathbf{p}$ is a left eigenvector of $\mathbf{N}$ with eigenvalue $(nl-1)$. The third of the equations above is simply $\mathbf{p\alpha} = -\mathbf{pN\alpha}$ in matrix form, so it follows that $\mathbf{p\alpha} = -(nl-1)\mathbf{p\alpha}$. This identity only holds if $\boldsymbol{\alpha}$ is in the nullspace of $\mathbf{p}$, such that $\mathbf{p\alpha} = 0$.

Treating a_i and α_i as the total and intrinsic selective advantage of the alleles implies a rather different interpretation of the Fundamental Theorem, and one which sits more comfortably with the modern gene-centric view of evolution. All of the equations that we have derived above hold, without further restriction, at any moment in time. Hence, were we to sample a population at successive times, the overall constraint $\sum_{i=1}^{s} p_i \alpha_i = 0$ that we derive in Box 2.3 would have to hold at any given census point. In a continuous-time framework, it follows that the constraint $\sum_{i=1}^{s} p_i(t) \alpha_i(t) = 0$ must naturally hold for all t. This conclusion leads to a beautifully simple interpretation of the Fundamental Theorem, because, upon differentiating the constraint $\sum_{i=1}^{s} p_i \alpha_i = 0$ using the product rule, we have the following constraint upon the evolutionary dynamics:

$$\sum_{i=1}^{s} \alpha_i \frac{dp_i}{dt} + \sum_{i=1}^{s} p_i \frac{d\alpha_i}{dt} = 0, \tag{2.6}$$

where the only new assumption we have made is that the intrinsic selective advantage (α_i) of every allele is differentiable with respect to time (t). As we explain below, each of the summations in Eq. 2.6 represents a flux in the intrinsic selective advantage of the alleles. For now, Eq. 2.6 may be rewritten using Eq. 2.5 to show that:

$$\sum_{i=1}^{s} \alpha_i \frac{dp_i}{dt} = -\sum_{i=1}^{s} p_i \frac{d\alpha_i}{dt} = \frac{V}{n} \tag{2.7}$$

where V is the additive genetic variance in fitness and n is the ploidy of the population. This shows that the first of the summations in Eq. 2.7, which is Fisher's '*rate of increase in fitness*' divided by n, is actually one of two equal and opposite fluxes that operate under natural selection. It is important to note that the balance between these two fluxes does not entail a state of equilibrium: it merely implies that some quantity is conserved. That quantity is the mean intrinsic selective advantage of all of the alleles, which is zero by definition. This, in essence, is the meaning of the constraint $\sum_{i=1}^{s} p_i \alpha_i = 0$ from which Eq. 2.6 is derived.

The first of the summations in Eq. 2.6, that is, $\sum_{i=1}^{s} \alpha_i (dp_i/dt)$, measures the positive flux that results directly from changes in allele frequency as alleles with positive intrinsic selective advantage spread, on average, through the population. This corresponds directly to Fisher's '*rate of increase in fitness*'. The second of the summations in Eq. 2.6, that is, $\sum_{i=1}^{s} p_i (d\alpha_i/dt)$, measures the negative flux that results directly from the reduction in intrinsic selective advantage that the same alleles experience as they become increasingly commonplace in the population. The balance between these two fluxes thereby captures exactly and completely the two essential characteristics of selective turnover in a population: the spread of selective advantage through the population, and the diminishment of selective advantage as it spreads. Fundamentally, this balance arises because evolution—understood to refer only to changes in allele frequency—represents a zero-sum situation, by virtue of the fact that changes in allele frequency must always sum to zero.

Although natural selection therefore gets the alleles nowhere in terms of their own intrinsic selective advantage, this does not mean to say that the properties of the individuals in the population do not improve in some meaningful way. On the contrary, an allele whose intrinsic selective advantage is positive will usually be adaptive in its effects upon individual survival or reproduction, and the spread of these alleles will therefore result in the spread of adaptation through the population.[5] We are therefore at liberty to conclude, with Fisher, that the additive genetic variance in fitness (V) characterizes exactly and completely the expected rate of adaptive improvement, or 'progress' in a population (Fisher, 1930).

2.2.3 Naming of the Fundamental Theorem

We may never know what Fisher himself understood by '*the rate of increase in fitness*', but a clue may be found in the fact that he felt able—apparently without hubris— to call his own theorem the 'Fundamental Theorem of natural selection', (Fisher, 1930). Fisher probably had in mind a connection with the Fundamental Theorem of calculus, which can be expressed as saying that, if a function has a continuous derivative, then the definite integral of that function is equal to the value of the function at the upper limit of the integral minus the value of the function at the lower limit of the integral. This would explain why Fisher chose to express the Fundamental Theorem not as a differential equation, but as the following equation involving differential forms (Fisher, 1958):

$$n \sum_{i=1}^{s} \alpha_i dp_i = V dt \tag{2.8}$$

which can be obtained straightforwardly from Eq. 2.5. Fisher called the left-hand side of Eq. 2.8 'the total increase in fitness', implying that he intended that Eq. 2.8 could indeed be integrated between two points in time. This supports our view that Fisher intended the overall constraint $\sum_{i=1}^{s} p_i \alpha_i = 0$ to hold at all times. Eqs. 2.6 and 2.7, which are the basis of our new interpretation of the Fundamental Theorem, follow immediately from the continuous imposition of this constraint.

In fact, the integral of $V dt$ amounts to what has been called the 'cumulative fitness flux' in recent fitness flux models of evolution derived from the asymptotics of diffusion equations borrowed from statistical physics (Mustonen and Lässig, 2009; Zhang et al., 2012). These models emphasise, as we have done here, that the important thing about natural selection is its flux, not what happens to population mean fitness, which is an irrelevance. It is difficult to know what Fisher's view would have been on this point, but a possible indication can be found in his own words of caution

---

[5] This need not always be the case, because the selective advantage of an allele is not necessarily aligned with the total reproduction of its bearer where kin selection or genetic conflict are concerned.

on the interpretation of the Fundamental Theorem: 'I have never, indeed, written about $\bar{W}$ [population mean fitness] and its relationships' (Fisher, 1941). We agree: Fisher wrote simply about a flux in the selective advantage of the alleles.

2.3 Fisher's quasi-conservation law of population mean fitness

Despite his protestations to the contrary, Fisher did make one very important and widely overlooked statement about population mean fitness (Frank, 2012). In a paragraph discussing competition for resources, he noted that population mean fitness, as measured by the Malthusian growth parameter of the population, could never be greatly different from zero (Fisher, 1930). Simply put, because no real population can grow indefinitely, and because any real population that contracts monotonically will soon be extinct, every successfully reproducing individual must tend to be replaced, on average, by approximately one successfully reproducing individual in the next generation (Frank, 2012). This simple argument leads to a 'quasi-conservation law' for population mean fitness, which Frank relates to the balance between the increasing adaptedness of individuals and the increasing adaptedness of their competitors under natural selection (Frank, 2012). As we have shown above, the balance between these two opposing effects is exact if they are modelled as fluxes in intrinsic selective advantage, whereas population mean fitness is still free to fluctuate—even if only a little—about a mean population growth rate of zero. These fluxes in the selective advantage of the alleles represent the dynamics of a zero-sum situation, and cannot themselves influence population mean fitness unless we assume arbitrarily that the effects of the alleles are constant (see also Mustonen and Lässig, 2009). Selective turnover is logically distinct from population growth (see also Ewens, 2011), and it is satisfying to note that there is, indeed, a clear and logical distinction between the evolutionary processes that govern population genetics, and the ecological processes that govern population dynamics.

2.4 Optimization and evolution

The most important use of our new interpretation of the Fundamental Theorem is in redefining Wright's metaphor of the adaptive landscape (Wright, 1932) so as to meet Fisher's entirely appropriate objection that population mean fitness cannot be modelled as a simple potential function (Fisher, 1941). Our revised model of the adaptive landscape will allow us to explain the optimizing tendency of natural selection in a rigorous manner, albeit without making the formal mathematical links that can be found in Grafen (2002) and Batty et al. (2013). Like many others who have adapted Wright's metaphor before us (reviewed in McGhee, 2007; Dietrich and Skipper Jr., 2012; Pigliucci, 2012), we take the 'vertical' axis of the landscape to represent individual fitness, and its 'horizontal' axes to represent phenotypic traits, rather than gene or genotype frequencies. However, we make some very specific distinctions in defining these terms, in order to relate them back to the fluxes in the selective advantage of the alleles that are the essential characteristic of natural selection.

2.4.1 *Redefining the adaptive landscape*

We consider first the vertical axis of the landscape. We have shown above that Fisher's Fundamental Theorem refers to an average flux in the intrinsic selective advantage of the alleles, defined as the additive contribution that the alleles make to their own rate of increase in the population. In general, the selective advantage of an allele depends not only upon the survival and reproduction of its own bearer, but also upon its social effects in the population. Social effects are irrelevant in most problems of interest in biomechanics, so it will usually be reasonable to use the vertical axis of the landscape to measure the expected rate of increase of the actual copy of an allele that is borne by an individual having the phenotypic properties associated with a given point on the horizontal axes. This will be linearly related to the relative fitness of the individual, defined as the number of successful gametes that the individual is expected to produce in its lifetime relative to the mean for all individuals in the population. It is therefore reasonable to take the vertical axis of the landscape to represent the relative fitness of an individual.

Our use of a relative measure of fitness follows directly from our derivation of the Fundamental Theorem above, and is consistent with the results of Grafen's 'formal Darwinism project', which also demonstrate the necessity of treating fitness as a relative quantity (Grafen, 2002, 2007, 2008; Batty et al., 2013; see also Frank, 2012). Relative fitness is expressed relative to the population mean, so it makes sense to incorporate population mean fitness as an explicit reference in our adaptive landscape metaphor. We may do this by treating population mean fitness as 'sea level', relative to which the height of the landscape is measured. Mathematically, sea level corresponds to the hyperplane satisfying the equation $w = 1$, where w is relative fitness. It follows that the actual mean fitness of the population is of no consequence to our adaptive landscape metaphor, because its absolute value is lost in the relativization of fitness. Representing population mean fitness as sea level also serves to emphasise the differences between our adaptive landscape metaphor and Wright's (Wright, 1932), which treated the height of the landscape itself as measuring population mean fitness.

Considering now the horizontal axes of the landscape, we take each of these to represent a quantitative phenotypic property of the individual that can be regarded as a performance objective for natural selection. By this, we mean any quantity whose increase would be expected to enhance the selective advantage of any allele conferring that increase, in the hypothetical case that such an increase could be effected without impacting performance in any other dimension. By the definition just given, the height of the landscape would be monotonically increasing along every horizontal axis, if every conceivable performance objective were represented by one of the horizontal axes. In reality, it is not usually possible to improve performance in one dimension without degrading performance in another. Hence, although selective advantage and relative fitness would be expected to increase monotonically in the direction of any horizontal axis, movement in this direction will usually be prevented by the trade-offs that exist between conflicting performance objectives. Consequently, the feasible domain of the fitness function, as defined by all of the constraints relating the different performance objectives, must be of lower dimension than the total number of performance objectives when all conceivable performance objectives are used as horizontal axes. This is because conflicting performance objectives are related by mathematical equalities, which mean that evolution has fewer degrees of freedom available to explore. We explore equality constraints of this sort in much greater detail in Chapter 3.

In the unlikely case that all conceivable performance objectives are represented by horizontal axes, the time-varying shape of the landscape will only depend upon how the different performance objectives combine to determine relative fitness. For example, in the special case that the performance objectives combine linearly, the shape of the landscape will depend only upon their relative weighting in the fitness function for a given organism in a given environment. In practice, we will only ever be able to work with a lower-dimensional representation of the landscape, in which case the time-varying shape of the landscape will also incorporate the trade-offs that exist between the performance objectives that are represented explicitly in the landscape and those which are not. In effect, the constraints defining the feasible domain of the problem are eliminated by incorporating them directly into the fitness function, which is a standard approach in constrained optimization (e.g. Boyd and Vandenberghe, 2004). This is important, because although we may not be able to say anything about how the different performance objectives combine to determine relative fitness, we may nevertheless be able to derive the physical constraints upon a problem from first principles. We can then use these constraint equations to define the subset of feasible solutions that might be favoured under natural selection, using the concepts of dominated and non-dominated solutions and Pareto optimality (Collette and Siarry, 2003; Boyd and Vandenberghe, 2004). We illustrate this procedure with respect to soaring performance in birds in Chapter 8.

2.4.2 *Drowning Mount Improbable*

The Fundamental Theorem predicts that alleles with a positive intrinsic selective advantage will tend to increase in frequency in the population (Section 2.2.2). Unfortunately, even perfect knowledge of changes in allele frequency is insufficient to predict exactly how the population will move across the landscape, because the genotypic combinations of alleles that are present in the population are liable to vary through time and the alleles need not combine linearly in their genotypic effects (see also Ewens, 2011). Nevertheless, the directionality of the flux in the intrinsic selective advantage of the alleles indicates that, at any given moment in time, an increasing proportion of the population may be expected to sit within the higher occupied regions of the landscape, as alleles with positive intrinsic selective advantage spread.

It is important to emphasize that this is only a general tendency, and that nothing specific can be said about where, exactly, on the landscape future individuals will sit. This will depend upon whatever new genotypic combinations of alleles are generated at reproduction, which cannot be forecast from changes in allele frequency (see Fisher, 1941).

The expected increase in the proportion of individuals that occupy the higher regions of the landscape will inevitably bring their own fitness closer to the population mean. Hence, because the height of the landscape above sea level represents individual fitness relative to the population mean, this conclusion implies that the landscape itself must subside under the influence of natural selection (see Rosenzweig, 1978, for a similar metaphor). In other words, selection itself alters the shape, or at least the height, of the adaptive landscape (see also Okasha, 2008). This process is expected to continue until the highest occupied peaks have been brought down to the sea. Hence, natural selection does not 'climb' Mount Improbable—she drowns it (cf. Dawkins, 1996).

By definition, those individuals that cling on near the waterline are no better off than their ancestors insofar as their relative fitness is concerned. They are, however, likely to be about as well adapted as they can be to the prevailing conditions, given the scope of the available genetic variation. Hence, given sufficient elapsed time, sufficient genetic variation, and sufficient environmental stability, individuals will indeed appear to act as if maximizing their fitness. This holds true even though the underlying directional tendency of natural selection relates not to individual fitness, but to the average flux in the intrinsic selective advantage of the alleles. Only when there is no remaining additive genetic variance in fitness, or when there is no room for further improvement will the process of natural selection cease. Grafen (2002) refers to these two situations as equilibria with 'no scope for selection' and 'no potential for selection', respectively. Metaphorically, the latter describes a situation in which the highest occupied points in the landscape have all been brought down to sea level. If evolution ever has an end point—or at least a point of stasis—then this is it.

2.5 Conclusions

On the view of natural selection that we have espoused in this chapter, species do not climb the adaptive landscape. Populations of individuals cling to its shores, part-drowning at the beachheads as they force down the land underfoot, until all but the loftiest peaks have sunk beneath the waves, and there is nothing left to conquer but the shifting sands on which the scattered remnant stands. It is for this reason, and no other, that individuals appear to act as if maximizing their fitness. It is possible that Darwin himself might have approved of this metaphor, paralleling as it does his own explanation of the formation of coral atolls (Darwin, 1842), but placing it in the context of his 'struggle for existence' (Darwin, 1859).

3

Constraint

3.1 Introduction

Every aspect of biology, from the energy budget of a hummingbird to the transpiration rate of a giant sequoia, is constrained by the physical basis of the natural world. Such constraints have been called 'universal constraints', because they arise from the universal laws of physics, or from invariant properties of materials and complex systems (Maynard Smith et al., 1985). Insofar as biology is ever a predictive science, it is the existence of physical constraints that most often makes it so. Maynard Smith et al. (1985) have contrasted 'universal constraints' with 'local constraints' arising from developmental processes intrinsic to the biology of particular taxa. Surprisingly, they have concluded that local developmental constraints are more easily identified and understood than are universal physical constraints. We contend that it is perfectly possible to identify and to understand universal physical constraints, and that to do so sheds light upon the interaction between selection and constraint more generally (see also Chapter 2). This chapter is about these universal physical constraints, and the role that they play in shaping the movements of animals.

In this chapter, we show how analytical techniques can be used to make theoretical predictions about the nature of physical constraint. We illustrate this with reference to the pendulum example that we introduced in Chapter 1, using the approach known as dimensional analysis to circumvent the need to derive the equations of motion explicitly (see e.g. Bridgman, 1922; Barenblatt, 2003). Next, we adapt the simple pendulum model to analyse the physical constraints upon the slow brachiating gait of gibbons. We then use similarly simplified physical models to explore the physical constraints upon walking and running. In the case of walking, we are able to formulate an explicit model to analyse the physical constraints in full. In the case of running, we are left simply with a prediction of the important dimensionless variables, which we then use to explore the constraints empirically. These analyses illustrate how to go about characterizing the operation of universal physical constraints. Simple theoretical analyses such as these should form the starting point of any empirical exploration of the evolutionary interaction between physical constraint, natural selection, and phylogeny. In other words, they ought really to form the starting point of any analysis in evolutionary biomechanics.

Evolutionary Biomechanics. Graham Taylor & Adrian Thomas.
© Graham Taylor & Adrian Thomas 2014. Published 2014 by Oxford University Press.

3.2 Dimensional analysis

Complex though animal movements may be, they are governed by Newtonian mechanics. This means that we may, in principle, write down a set of differential equations with the same dynamics as the organism itself. This makes the physical constraints upon animal locomotion susceptible to the kind of description that would be inconceivable for most of the other behaviours that animals exhibit. In order to say something interesting about the nature of these constraints, we do not have to be able to write out the underlying differential equations in full. This is because the very existence of such equations allows us to use dimensional analysis to identify the most fundamental physical constraints upon the problem. Dimensional analysis is an especially useful tool in problems whose complexity defies literal analysis. Such is the nature of biomechanics.

The essence of a dimensional analysis is to reduce the complexity of a problem by manipulating all of the important dimensional variables, such as mass, length, time, velocity, acceleration, or force, into a smaller number of mathematically independent dimensionless variables. For example, the ratio of one length measurement to another is obviously dimensionless, but so too is the square of a velocity divided by the product of a length and an acceleration, because the units of length and time cancel. The central result of dimensional analysis is formalized in the Pi theorem (Buckingham, 1914), which states that any unique physical relationship connecting n different variables can be rewritten as a relationship between k fewer dimensionless variables, where k is the number of the original variables having independent dimensions (see also Bridgman, 1922; Barenblatt, 2003). This statement requires some unpacking, and is best explained by example.

The first step in carrying out a dimensional analysis is to identify which variables are important to the problem in hand. Conceptually, these are the variables that would have to appear in any reasonable mathematical model of the system (Bridgman, 1922). Suppose, for example, that we want to analyse how the period of a clock pendulum is constrained by its physics. In order to identify the important variables, our first step will be to conceive a simplified model of the system. The classical idealization of a pendulum is known as a 'simple pendulum' (Figure 3.1), and represents the system as a point mass hanging from a rigid massless arm that is free to rotate about a fixed, frictionless pivot under the influence of gravity (e.g. Baker and Blackburn, 2005). The pendulum swings freely and so has zero velocity when it reaches the top of its swing. Because we are neglecting friction effects, the only variables that can possibly affect the period (T) of the pendulum are its mass (m), its length (l), its angle at the top of each swing (Θ) and the acceleration it experiences due to gravity (g). The dimensions of these variables are listed in Table 3.1 in terms of the fundamental units of mass (M), length (L) and time (T).

The dimension of any mechanical variable is always a power-law monomial of the three fundamental mechanical units, M, L, and T, meaning that it is of the general form $M^x L^y T^z$, where x, y, and z are numerical exponents. Additional fundamental units would be needed to describe a thermodynamic or electromagnetic system, but

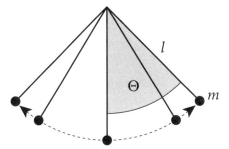

Figure 3.1 The simple pendulum model. The pendulum is modelled as a point mass (m) hanging by a rigid massless arm of length l from a fixed, frictionless pivot about which it is free to rotate under the influence of gravity. The pendulum reaches its maximum angle with respect to the vertical (Θ) at the top of each swing.

with only three fundamental units to play with, it is obvious that no more than three dimensional variables can ever have independent dimensions. For example, the dimension of the mass of a pendulum (M) is clearly independent of both the dimension of its period (T) and the dimension of its length (L), because each involves a different fundamental unit. However, the dimension of gravitational acceleration (LT^{-2}) obviously depends upon the dimensions of the preceding variables, because it involves the same fundamental units as do the dimensions of pendulum length and period. Hence, of the $n = 5$ variables that we have identified as being possibly important to pendulum dynamics in Table 3.1, there are $k = 3$ with independent dimensions.

As $n - k = 2$, the Pi theorem states that we can reduce the unknown dimensional relationship connecting the five variables to a dimensionless one involving only two dimensionless variables. The angle of the pendulum at the top of its swing (Θ) is already dimensionless, so may straightforwardly be used as the first dimensionless variable. The second dimensionless variable has to be independent of the first, and must therefore be formed as a power-law monomial of the four remaining dimensional variables. In other words, it is of the general form $T^w m^x l^y g^z$, where w,

Table 3.1 Variables to be considered in a dimensional analysis of a simple pendulum

Variable	Symbol	Dimension
pendulum period	T	T
pendulum length	l	L
pendulum mass	m	M
gravitational acceleration	g	LT^{-2}
angle at top of swing	Θ	dimensionless

x, y and z are numerical exponents for which we must solve. The dimension of this power-law monomial is $T^w M^x L^y (LT^{-2})^z$. We know that the exponents of each fundamental unit must sum to zero if it is to be dimensionless, so upon collecting terms, we have the following simultaneous equations to solve:

$$
\begin{aligned}
\text{for M}: & \quad x = 0 \\
\text{for L}: & \quad y + z = 0 \\
\text{for T}: & \quad w - 2z = 0.
\end{aligned}
\tag{3.1}
$$

Because $x = 0$, we can see immediately that the exponent of the pendulum's mass is zero. It follows that the period of a simple pendulum must be independent of its mass. This is a remarkable result to have arrived at, considering how few assumptions we have made.

These simultaneous equations still leave us with one unknown; but if we arbitrarily set $w = 1$, then we can solve for the remaining exponents as $y = -1/2$ and $z = 1/2$. Substituting these exponents back into $T^w m^x l^y g^z$, we are able to form the dimensionless variable $T\sqrt{g/l}$, which we will call the 'dimensionless period'. We would, of course, have arrived at a different form for this dimensionless variable if we had selected a different value for w, but this is immaterial to the argument, because all that the Pi theorem tells us is that our second dimensionless variable is some undetermined function of the first. It follows that we may immediately write the equation

$$
T\sqrt{\frac{g}{l}} = \Phi(\Theta)
\tag{3.2}
$$

where Φ is some undetermined function of Θ. This function can only be determined empirically, or with the aid of an explicit analytical model.

In principle, we cannot tell from Eq. 3.2 whether Φ is an increasing or decreasing function of Θ at any given point. The example of a simple pendulum is unusual, however, because the undetermined function depends upon only one variable. Consequently, the function $\Phi(\Theta)$ must simplify to a constant if the swing angle Θ is held constant. It follows that the dimensionless period $T\sqrt{g/l}$ must be the same for any simple pendulum swinging from a given angle Θ. As gravity is more or less constant on Earth, this implies that pendulum period is constrained to scale as $\sqrt{l}$, which is the result that we found empirically for clock pendulums in Chapter 1. Eq. 3.2 therefore expresses a universal physical constraint upon the way that pendulums scale. Because it involves a mathematical equality, we follow the mathematical optimization literature in referring to it as an 'equality constraint' (e.g. Boyd and Vandenberghe, 2004). This result is unusual in its elegance, but as we now show, the same technique can be used to reach some useful conclusions about physical constraints in biomechanics. We begin by considering swinging gaits.

3.3 Swinging gaits

3.3.1 Theoretical constraints upon pendular brachiation

Gibbons (Hylobatidae) have evolved to use an unusual mode of locomotion, called brachiation, to swing hand over hand through the forest canopy (Figure 3.2). Gibbons ricochet from one branch to the next when moving quickly, but during slow locomotion they keep one hand in contact with the branch at all times. During this continuous-contact gait, the loss of energy through collisions with the branches is small, and the dynamics involve a pendulum-like interchange of kinetic and potential energy (Swartz, 1989; Bertram, 2004). The mass of a gibbon is obviously not concentrated at a single point, as it is in the simple pendulum model, but it can be shown that the equation of motion for a pendulum with distributed mass is the same as the equation of motion for a simple pendulum with a shorter arm.[1] Hence, if we are willing to consider only the natural dynamics, and to neglect friction forces and aerodynamic drag, then the simple pendulum model may be a serviceable approximation to pendular brachiation in gibbons.

Although Eq. 3.2 could be used to model the physical constraints upon pendular brachiation, it is not a useful way of expressing the constraints, because swing period is of no obvious biological importance. On the other hand, we can convert Eq. 3.2 into

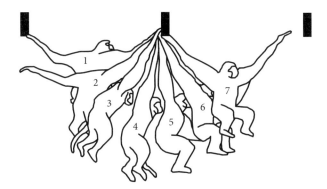

Figure 3.2 Brachiating gibbon, showing the animal in seven successive positions at constant time intervals. Although brachiation is a complex motion, its dynamics involve a pendulum-like interchange between kinetic and gravitational-potential energy. This is visible in the spacing of the successive positions, which increases towards the bottom of the swing as the kinetic energy of the motion increases, and decreases towards the top of the swing as kinetic energy is exchanged for gravitational-potential energy. The same is true of a swinging pendulum. Redrawn from Bertram (2004). Reproduced with permission of John Wiley & Sons.

---

[1] The length of the equivalent simple pendulum is $l = I/(md)$, where m is the total mass, I is the moment of inertia about the pivot, and d is the distance from the pivot to the centre of mass (e.g. Baker and Blackburn, 2005).

Table 3.2 Variables to be considered in a dimensional analysis of pendular brachiation

Variable	Symbol	Dimension
pendulum length	l	L
stride length	λ	L
body mass	m	M
gravitational acceleration	g	LT^{-2}
speed of brachiation	U	LT^{-1}

a much more useful form by noting that swing period is equal to stride length divided by mean forward speed. This will allow us to examine the physical constraints upon the speed of pendular brachiation. Having so introduced stride length to the problem, we need also to note that the angle at the top of each swing is trigonometrically related to pendulum length and stride length. When listing the variables that are important to brachiation, we may therefore replace the pendulum period and angle at the top of a swing with speed and stride length. Hence, the speed of brachiation (U) might in principle depend upon the mass of the body (m), the stride length (λ), the gravitational acceleration (g), and the pendulum length (l). The dimensions of these variables are listed in Table 3.2.

Of the $n = 5$ variables, there are $k = 3$ with independent dimensions. Hence, as $n - k = 2$, the Pi theorem states that we can reduce the dimensional relationship connecting the five variables to a dimensionless one involving only two dimensionless variables. It is clear by inspection that we can form the first of these by combining stride length λ and pendulum length l as λ/l. We will call this the 'relative stride length', by analogy with the walking literature (Alexander, 1976). The second dimensionless variable must be independent of the first, and so cannot include both stride length and pendulum length. Choosing arbitrarily not to include stride length in the second dimensionless variable, we must therefore combine the four remaining dimensional variables, l, m, g, and U, into a dimensionless variable of the form $l^w m^x g^y U^z$, where w, x, y and z are numerical exponents for which we must solve. The dimension of this power law monomial is $L^w M^x \left(LT^{-2}\right)^y \left(LT^{-1}\right)^z$, where the exponents of each fundamental unit must sum to zero. Collecting terms, we therefore have the following simultaneous equations to solve:

$$
\begin{aligned}
\text{for } M : &\quad x = 0 \\
\text{for } L : &\quad w + y + z = 0 \\
\text{for } T : &\quad -2y - z = 0.
\end{aligned}
\tag{3.3}
$$

We can see by inspection that $x = 0$, so the exponent of body mass in the second dimensionless product is zero. We may set the value of z arbitrarily, and it will prove convenient to set $z = 2$. We may then solve for the other exponents as $y = -1$ and

$w = -1$. Substituting these values back into $l^w m^x g^y U^z$ yields the conventional dimensionless variable $U^2/(gl)$, which we will call the 'Froude number' by analogy with the walking literature (Alexander, 1976).

It follows that a dimensional relationship describing the dynamics of pendular brachiation can be rewritten as a dimensionless one involving only the Froude number and relative stride length. The dynamics of pendular brachiation are therefore expected to be independent of body mass, which does not feature in either the Froude number or the relative stride length. We may summarize this relationship this by writing

$$\frac{U^2}{gl} = \Phi \left(\frac{\lambda}{l} \right) \tag{3.4}$$

where Φ is some new and undetermined function. This equation implies that the Froude number ($U^2/(gl)$) of an animal brachiating in a pendulum-like fashion depends only upon its relative stride length (λ/l). Eq. 3.4 does not say anything about the form of the relationship between these variables, so we cannot tell from Eq. 3.4 alone whether the Froude number is an increasing or decreasing function of relative stride length at any given point (although it is obvious upon energetic grounds that the function must be an increasing one). All that we know from Eq. 3.4 is that the Froude number and relative stride length are uniquely constrained by an equation involving only those two variables (Eq. 3.4). We identify the form of that constraint empirically in the next section. It is important to note that this equality constraint only applies during continuous-contact gaits. One way to escape from this particular constraint is to change to a less-constrained ricochetal style of brachiation involving an aerial phase (Bertram et al., 1999; Bertram and Chang, 2001).

3.3.2 Empirical constraints upon brachiation

Postulating a constraint on the basis of a simplified physical model is one thing, but showing that it is important in practice requires us to test our prediction against empirical data. Chang et al. (2000) studied a white-handed gibbon, *Hylobates lar*, swinging between handholds whose spacing could be varied experimentally. Although we do not know the length of the equivalent simple pendulum for this individual, we do know its arm length, and can use this as a proxy for pendulum length in calculating relative stride length and the Froude number. The data from Chang et al. (2000) do indeed show a strong positive relationship between log Froude number and log relative stride length (Figure 3.3a), where the variables have been log-transformed because although the function relating them is unknown, the known scaling of a simple pendulum leads us to expect that it might be approximated as a power-law relationship. Although the empirical relationship between log Froude number and log relative stride length is by no means perfect, the strength ($R^2 = 0.92$) and significance ($F_{1,83} = 940$; $p < 0.0005$) of its linear component is consistent with

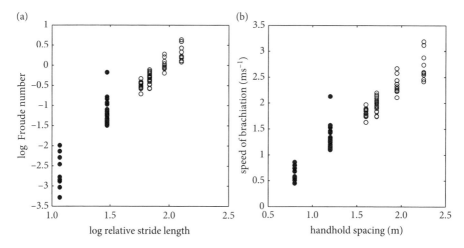

Figure 3.3 (a) Plot of log Froude number against log relative stride length for a brachiating white-handed gibbon *Hylobates lar* (Chang et al., 2000). Dimensional analysis predicts that Froude number and relative stride length will be physically related if the swing dynamics are pendular. Filled circles denote continuous-contact gaits; open circles denote gaits with an aerial phase. (b) Plot of brachiation speed against handhold spacing. The speed of forward motion is linearly correlated with handhold spacing, as expected if the swing dynamics are pendular. This is consistent with the hypothesis that the spacing of available handholds constrains the speed of brachiation. Original data plotted by kind permission of Young-Hui Chang.

our theoretical prediction that these two variables should be subject to an equality constraint.[2]

This equality constraint has some important consequences. For example, we expect from Eq. 3.4 that the speed of a gibbon using pendular brachiation will vary as a function of its stride length, which in turn means that the speed of locomotion should be constrained by the spacing of the available handholds. In fact, if the dynamics were truly pendular, such that the stride period was approximately constant, then we would expect to see a linear relationship between the speed of brachiation and the spacing of the available handholds. The data from Chang et al. demonstrate that there is indeed a positive and approximately linear relationship between speed and handhold spacing for an individual gibbon (Figure 3.3). This is consistent with the theoretical prediction that the distance between available handholds constrains the speed of pendular brachiation, although the same relationship also appears to

---

[2] Care needs to be taken when assessing the strength of relationship between dimensionless variables. Froude number and relative stride length both have pendulum length in their denominator, so any variation in pendulum length will result in covariance between the two dimensionless variables, regardless of whether they are physically related (see also Nee et al., 2005). This is not an issue here, because the data are drawn from a single individual and are therefore assumed to exhibit no variation in pendulum length.

hold in gaits with an aerial phase (Figure 3.3). Pendular dynamics can be highly efficient, but they are also highly constrained. This may be one reason why brachiating gibbons switch from a continuous-contact gait to a less-constrained ricochetal gait when they want to move quickly (Bertram et al., 1999; Bertram and Chang, 2001).

3.4 Walking gaits

3.4.1 Theoretical constraints upon walking

We now consider the more complex case of walking. Like the dynamics of brachiation, the dynamics of walking are characterised by a periodic transfer between kinetic and gravitational-potential energy (Cavagna et al., 1977). The simplest possible model of this transfer of energy represents the animal as an inverted simple pendulum (e.g. Alexander, 2003). This idealization treats the body as a point mass supported by a rigid massless leg, which is held in contact with the ground without slipping (Figure 3.4). Real animals do not conform to this idealization, of course, but the equation of motion for an inverted real pendulum with distributed mass is the same as the equation of motion for an equivalent simple pendulum with a slightly shorter leg (e.g. Baker and Blackburn, 2005).

By analogy with the simple pendulum model that we developed for brachiation, the mean forward speed of walking (U) might in principle depend upon the mass of the body (m), the stride length (λ), the pendulum length (l), and the gravitational acceleration (g). However, whereas a swinging pendulum has no kinetic energy when it reaches its highest point, an inverted pendulum must have non-zero kinetic energy at its highest point in order to move the mass past the vertical. The additional energy that is required to move past the vertical adds an extra degree of freedom to the problem, which we can deal with by including the angular velocity at mid-stance (ω)

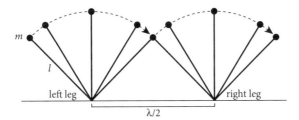

Figure 3.4 Inverted pendulum model of walking. The body is modelled as a point mass m supported by a rigid leg of length l, assumed to have no mass. The system moves freely under the influence of gravity through a stride of length λ. The recovery strokes of the legs are not modelled: each leg is instead assumed to appear in the correct position at the beginning of the stance phase. The inverted pendulum model is evidently not intended as a literal physical model of walking, but instead serves to supply the relevant variables for the dimensional analysis.

Table 3.3 Variables to be considered in a dimensional analysis of the inverted pendulum model of walking

Variable	Symbol	Dimension
pendulum length	l	L
stride length	λ	L
body mass	m	M
gravitational acceleration	g	LT^{-2}
walking speed	U	LT^{-1}
angular velocity mid-stance	ω	T^{-1}

in our model. The dimensions of all six variables are listed in Table 3.3. Of these, $k = 3$ have independent dimensions. This is easiest to see for l, m and ω, which each have a dimension involving a different fundamental unit. The Pi theorem therefore states that we can reduce the dimensional relationship between the $n = 6$ variables to a dimensionless one involving only $n - k = 3$ dimensionless variables. There are numerous different ways in which the dimensional variables could be combined to form three dimensionless variables, but it is clear by inspection that we have all the variables we need to use the same two dimensionless variables as we derived in the swinging pendulum model of brachiation. These are the Froude number ($U^2/(gl)$), and the relative stride length (λ/l).

The third dimensionless variable must be independent of the Froude number and relative stride length. We can ensure this by choosing not to include stride length or walking speed in this third dimensionless product, which must therefore combine the remaining four variables, l, m, g, and ω, into a dimensionless variable of the form $l^w m^x g^y \omega^z$, where w, x, y, and z are numerical exponents for which we must now solve. The dimension of this power law monomial is $L^w M^x \left(LT^{-2}\right)^y \left(T^{-1}\right)^z$, where the exponents of each fundamental unit must sum to zero. Collecting terms, we therefore have the following simultaneous equations to solve:

$$\begin{aligned} &\text{for } M: &x = 0 \\ &\text{for } L: &w + y = 0 \\ &\text{for } T: &-2y - z = 0. \end{aligned} \qquad (3.5)$$

We can see by inspection that $x = 0$. Making the arbitrary decision to set $z = 2$, we may now solve for the other exponents as $y = -1$ and $w = 1$. Substituting these values back into $l^w m^x g^y \omega^z$ yields the dimensionless variable $l\omega^2/g$. Alexander (2003) has occasionally referred to this quantity as a 'Froude number'. However, it is free to vary independently of the usual Froude number ($U^2/(gl)$), and we will instead call it the 'dimensionless vault speed', because it reflects the angular speed of the body as it vaults over the leg mid-stance. Physically, dimensionless vault speed measures the ratio of centripetal acceleration to gravitational acceleration at the top of the stride, which becomes relevant again later in exploring the gait transition from walking to running.

Mass does not feature in any of the three dimensionless variables that we have identified, so it follows that the mass of the body is predicted to have no effect on the dynamics of inverted pendulum walking. This is the same result as we found earlier for swinging pendulums, and a physical explanation of this is that mass affects kinetic and gravitational-potential energy in direct proportion. It follows that changes in mass do not affect the dynamics of the conservative transfer of energy between these two forms. Surprising as it may look, it is not unexpected, therefore, that women of certain African tribes walk quite normally while carrying loads up to 70% of body weight on their heads (Maloly et al., 1986). Carrying a large weight on the head increases the length of the equivalent inverted simple pendulum by changing the total mass distribution. Given that efficient walking animals tend to have long legs, this increase in effective pendulum length might go some way to explaining the efficiency with which humans are able to carry heavy loads on the head.

We have shown that the dimensional relationship describing the dynamics of inverted pendulum walking may be rewritten as a dimensionless one between the Froude number, relative stride length, and dimensionless frequency. We may summarize this by writing

$$\frac{U^2}{gl} = \Phi\left(\frac{\lambda}{l}, \frac{l\omega^2}{g}\right) \tag{3.6}$$

where Φ is some new and undetermined function which defines an equality constraint upon inverted pendulum walking. Once again, Eq. 3.6 leaves undetermined whether the function Φ is an increasing or decreasing function of its arguments at any given point. Alexander (1976) concluded, via an informal dimensional analysis, that the Froude number and relative stride length should be physically related in walking and running. However, our dimensional analysis shows that the Froude number in walking gaits depends not only upon relative stride length, but also upon dimensionless vault speed. In principle, the relationship between these three variables may be determined either empirically or with the aid of an explicit analytical model.

Surprisingly, the inverted pendulum model of walking does not appear to have been solved analytically before. The solution is outlined in Box 3.1 and conforms to the general form of the dimensionless relationship in Eq. 3.6, although the undetermined function Φ turns out to be rather complicated. The mathematical details of the model are not important for our purposes, but we graph it in Figure 3.5a to illustrate the equality constraint relating the Froude number, relative stride length, and dimensionless vault speed in inverted pendulum walking. The biomechanical significance of this graph is that any animal walking exactly like an inverted pendulum would be constrained by its own dynamics to lie somewhere upon the surface that this equality constraint defines. Real animals do not conform exactly to this model, of course, but they cannot deviate too far from it if they walk by vaulting over stiff legs.

Box 3.1 Analytical solution of the inverted pendulum model of walking.

The sum of the kinetic and gravitational-potential energy of a simple inverted pendulum is

$$\frac{1}{2}m\left(l\frac{d\theta}{dt}\right)^2 - mgl\left(1-\cos\theta\right)$$

where m is body mass, l is pendulum length, θ is the angle of the pendulum from the vertical, t is time, and g is gravitational acceleration. Because the system is conservative, we may therefore write the energy conservation equation

$$\frac{1}{2}m\left(l\omega\right)^2 = \frac{1}{2}m\left(l\frac{d\theta}{dt}\right)^2 - mgl\left(1-\cos\theta\right)$$

where ω is the angular velocity at $\theta = 0$. This can be rearranged to yield

$$\frac{d\theta}{dt} = \pm\left(\omega^2 + \frac{2g}{l}\left(1-\cos\theta\right)\right)^{\frac{1}{2}}$$

where the plus (minus) sign indicates rotation in the positive (negative) sense of θ. Stride period (T) can be calculated by inverting this equation and integrating $dt/d\theta$ between appropriate limits. Defining λ as stride length, the movement of the pendulum from $\theta = 0$ to $\theta = \arcsin \lambda/(4l)$ takes a quarter of a stride cycle. We may therefore write

$$T = 4\int_0^{\arcsin\frac{\lambda}{4l}}\left(\omega^2 + \frac{2g}{l}\left(1-\cos\theta\right)\right)^{-\frac{1}{2}}d\theta$$

which can be evaluated in terms of an elliptic integral of the first kind (F):

$$T = \frac{8}{\omega}F\left(\frac{1}{2}\arcsin\frac{\lambda}{4l}\,\middle|-4\frac{g}{l\omega^2}\right).$$

Using the identity $U = \lambda/T$, we can now write an equation for walking speed (U):

$$U = \frac{\lambda\omega}{8}\left[F\left(\frac{1}{2}\arcsin\frac{\lambda}{4l}\,\middle|-4\frac{g}{l\omega^2}\right)\right]^{-1}.$$

This can be rewritten as an equation for the Froude number by squaring it and dividing through by gravitational acceleration and pendulum length:

$$\frac{U^2}{gl} = \frac{1}{64}\left(\frac{l\omega^2}{g}\right)\left(\frac{\lambda}{l}\right)^2\left[F\left(\frac{1}{2}\arcsin\frac{\lambda}{4l}\,\middle|-4\frac{g}{l\omega^2}\right)\right]^{-2},$$

although it is interesting to note that putting the equation in dimensionless form complicates it considerably. The resulting dimensionless equation specifies the equality constraint upon inverted pendulum walking in terms of the Froude number ($U^2/(gl)$), relative stride length (λ/l), and dimensionless vault speed ($l\omega^2/g$), and is graphed in Figure 3.5a.

Box 3.2 Limits of the inverted pendulum model of walking.

Clearly, dimensionless vault speed and relative stride length must both be greater than zero for forward motion to occur. Moreover, the leg will only remain in contact with the ground if the component of gravitational acceleration acting along it, $g \cos \theta$, is sufficient to provide the necessary centripetal acceleration, $l(d\theta/dt)^2$. Making the substitution $l(d\theta/dt)^2 \leq g \cos \theta$ into the third of the equations in Box 3.1 and rearranging yields

$$\frac{l\omega^2}{g} \leq -2 + 3 \cos \theta.$$

The right-hand side of this inequality is smallest at maximum leg excursion, which occurs when $\theta = \arcsin \lambda/(4l)$. Making this substitution, the upper limit on dimensionless vault speed is given by

$$\frac{l\omega^2}{g} \leq -2 + 3 \cos \left(\arcsin \frac{\lambda}{4l} \right) = -2 + 3 \sqrt{1 - \frac{1}{16} \left(\frac{\lambda}{l} \right)^2}.$$

Slipping will occur if the ratio of the horizontal and vertical centripetal force components, $\tan \theta$, exceeds the static coefficient of friction (μ). The ratio of these components is greatest at maximum leg excursion, so the upper limit on relative stride length is

$$\frac{\lambda}{l} \leq 4 \sin (\arctan \mu) = \frac{4\mu}{\sqrt{1 + \mu^2}}.$$

These upper limits on dimensionless vault speed and relative stride length are the inequality constraints upon inverted pendulum walking.

So far, we have simply assumed that the leg is held in contact with the ground without slipping. In order to satisfy this assumption, the component of gravitational acceleration acting along the leg must be sufficient to provide the centripetal acceleration needed to keep the mass of the body moving in a circular arc. This allows us to place an upper limit upon dimensionless vault speed (Box 3.2). Furthermore, the ratio of horizontal to vertical centripetal force can never exceed the static coefficient of friction, because the horizontal friction force available to resist sliding is equal to the vertical centripetal force times the static coefficient of friction with the substrate. This allows us to place an upper limit upon relative stride length for any given static coefficient of friction (Box 3.2). These limits are plotted in Figure 3.5a and are similar to those calculated numerically by Usherwood (2005). Both limits are specified by a mathematical inequality (Box 3.2), and we therefore follow the mathematical optimization literature in referring to them as 'inequality constraints' (e.g. Boyd and Vandenberghe, 2004).

The equality constraint relating the Froude number to relative stride length and dimensionless vault speed sets the contours of the surface in Figure 3.5a; the inequality constraints upon dimensionless vault speed and relative stride length set its boundaries. The surface in Figure 3.5a therefore represents the design space that animals can explore if they make use of inverted pendulum walking. In general, equality constraints define a hypersurface which organisms are free to explore behaviourally or through evolution, whilst inequality constraints delimit the boundaries of that exploration.

3.4.2 Empirical constraints upon walking

Figure 3.5b plots combinations of Froude numbers and relative stride lengths for 11 species of walking and running mammal, taking hip height as a proxy for pendulum length in the inverted pendulum model (Alexander and Jayes, 1983). Alexander and Jayes did not measure dimensionless vault speed, but the first of the inequality constraints in Box 3.2 can be used to predict an upper limit on the Froude number for a given relative stride length. This limit contains all of the empirical data points for walking gaits, and predicts reasonably well the transition to gaits with an aerial phase. Hence, walking does indeed appear to transition to running at speeds at which gravitational acceleration is only just sufficient to provide the centripetal acceleration needed to keep the leg in contact with the ground. In contrast, the upper limit on relative stride length set by the second of the inequality constraints in Box 3.2 does not appear to be an important constraint in the data plotted in Figure 3.5B. This limit is an increasing function of the static coefficient of friction (Box 3.2), so it is only likely to act as an important constraint on the most slippery surfaces. Humans take shorter steps when walking on ice, for example, but do not otherwise tend to modulate stride length according to substrate.

The measurements of Froude number and relative stride length in Figure 3.5b are overlain by isolines of the predicted dimensionless vault speed. These are simply the contours of the surface drawn in Figure 3.5a, which we have projected onto the plane defined by Froude number and relative stride length. A notional line of best fit through the data would evidently cross several of these isolines, indicating that higher relative stride lengths are associated with higher dimensionless vault speeds. In fact, increasing relative stride length without changing dimensionless vault speed (i.e. moving along the isolines) has relatively little effect upon Froude number, except at unrealistically low dimensionless vault speeds (Figure 3.5b). This reflects the fact that the inverted pendulum model (Box 3.1) predicts that the Froude number should be almost directly proportional to dimensionless vault speed, with only a small amount of curvature due to relative stride length (Figure 3.5a). Mathematically, this is because the elliptic integral in the equation (Box 3.1) is nearly linear in relative stride length, except at very small dimensionless vault speeds. This all but cancels the effects of relative stride length elsewhere in the equation.

If this result seems surprising, then try seeing how quickly you can walk at a fixed stride length by increasing the speed with which you propel yourself forward on each stride. It will quickly become apparent that speed can easily be varied by walking more or less energetically in the same set of footprints. The reason why relative stride length is almost incidental to Froude number is that it influences the Froude number of an inverted pendulum through its effects upon the interchange of kinetic and gravitational-potential energy (Figure 3.4). The magnitude of this interchange is

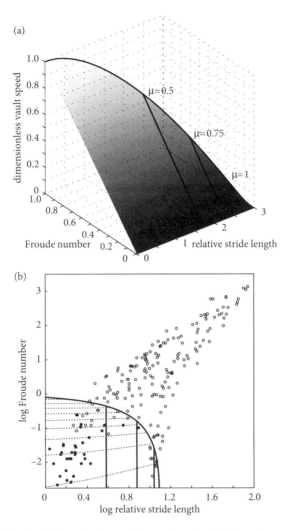

Figure 3.5 (a) Graph of dimensionless vault speed against the Froude number and relative stride length for the inverted pendulum model of walking (Box 3.1). The curved line denotes the upper limit on dimensionless vault speed, while the lines marked $\mu = 0.5$, $\mu = 0.75$, and $\mu = 1$ each denote the upper limit on relative stride length for a different value of the static coefficient of friction (Box 3.2). As context for these values, US courts of law recognize $\mu = 0.5$ as the minimum for slip-resistant surfaces. (b) Plot of log Froude number against log relative stride length for 11 species of mammals (Alexander and Jayes, 1983). Filled circles denote walking; open circles denote running. The straight solid lines in (b) correspond directly to the limits on relative stride length for the three values of μ given in (a). The solid curved line represents the limit beyond which gravitational acceleration is insufficient to keep the leg in contact with the ground at all times. We would therefore expect any point beyond this limit to represent a running gait (filled circles), which is indeed the case. The dotted lines are the contours of the surface in (a), which are isolines of dimensionless vault speed. Original data plotted by kind permission of R. McNeill Alexander.

small in comparison with the total kinetic energy of the motion, except at very low dimensionless vault speeds, so it follows that the effects of relative stride length are also comparatively small. The empirical relationship that Alexander and Jayes (1983) identified between Froude number and relative stride length cannot, therefore, be explained as a direct or inevitable consequence of the dynamics of inverted pendulum walking. It is not the result of physical constraint—at least not according to this simplest of physical models.

Froude number and relative stride length both have pendulum length in their denominator, so any variation in pendulum length is bound to introduce covariance between Froude number and relative stride length. This covariance is not sufficient to explain the empirical relationship between Froude number and relative stride length, however, because we would expect Froude number to vary linearly with relative stride length if this were the whole story. In fact, their empirical relationship is more nearly cubic. We are left with the conclusion that the relationship between Froude number and relative stride length is likely to be the result of adaptation, rather than being a statistical artifact or the result of physical constraint. It is certainly possible to think of adaptive reasons why Froude number and relative stride length might be related empirically. Perhaps, as Alexander and Jayes (1983) suggest, this is because of selection to minimize the cost of transport. Alternatively, it may be because the duration of the power stroke, and hence the dimensionless vault speed, can be increased by increasing relative stride length. Unfortunately, neither of these possibilities can be accommodated within the inverted pendulum model of walking, which assumes no energetic losses and an impulsive start to the motion.

3.5 Running gaits

3.5.1 Theoretical constraints upon running

Real animals do not move stiffly—even when walking (Geyer et al., 2006), and especially when running. The dynamics of running are characterised by a periodic transfer between kinetic and elastic-potential energy (Cavagna et al., 1977). The simplest model of this transfer (e.g. Alexander, 2003) represents the animal as a planar mass-spring system, in which a linear spring of stiffness K is incorporated into the leg of an inverted simple pendulum (Figure 3.6). The set of variables that we need to consider is listed in Table 3.4, and includes the angular amplitude of the stance phase of each step (Θ). Angular amplitude plays a critical role in determining the take-off angle at the start of the aerial phase, and is therefore a more meaningful parameter to include than the angular velocity at mid-stance, which it replaces in the list of variables (cf. Table 3.3). Note that although angular amplitude is trigonometrically related to pendulum length and stride length in walking, the occurrence of an aerial phase means that the angular amplitude is free to vary independently in running. Of the $n = 7$ variables listed in Table 3.4, $k = 3$ have independent

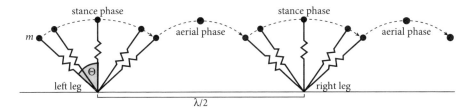

Figure 3.6 Planar mass-spring model of running. The body is modelled as a point mass m supported by a compliant leg of unloaded length l, subject to compression under the influence of gravity. During the stance phase, the leg sweeps through an angle 2Θ. Every stance phase is followed by a ballistic aerial phase. The total stride length is denoted λ. The recovery strokes of the legs are not modelled: each leg is instead assumed to appear in the correct position at the beginning of the stance phase. Like the inverted pendulum model of walking, the planar mass-spring model is not intended as a literal physical model of running, but instead serves to supply the relevant variables for the dimensional analysis.

Table 3.4 Variables to be considered in a dimensional analysis of the planar mass-spring model of running

Variable	Symbol	Dimension
leg length	l	L
stride length	λ	L
body mass	m	M
gravitational acceleration	g	LT^{-2}
running speed	U	LT^{-1}
leg stiffness	K	MT^{-2}
angular amplitude	Θ	dimensionless

dimensions. Hence, as $n - k = 4$, the Pi theorem states that we can rewrite the dimensional relationship between the variables as an equation involving four dimensionless variables.

It is clear from the list of variables (Table 3.4) that we may use the angular amplitude (Θ), Froude number ($U^2/(gl)$), and relative stride length (λ/l) as our first three dimensionless variables. This leaves us to write a fourth dimensionless product involving the leg stiffness (K). We can ensure that this is independent of the other three dimensionless variables by choosing not to use stride length, running speed, or angular amplitude in forming the product. The fourth dimensionless variable must therefore combine the remaining dimensional variables, K, l, m, and g, into a single product of the form $K^w l^x m^y g^z$, where w, x, y, and z are numerical exponents for which we must solve. The dimension of this power law monomial is $\left(MT^{-2}\right)^w L^x M^y \left(LT^{-2}\right)^z$, where the exponents of each fundamental unit must sum to zero. Collecting terms, we therefore have the following simultaneous equations to solve:

$$
\begin{array}{ll}
\text{for } M: & w + y = 0 \\
\text{for } L: & x + z = 0 \\
\text{for } T: & -2w - 2z = 0.
\end{array}
\tag{3.7}
$$

Making the arbitrary decision to set $w = 1$, it follows that $y = -1$, $z = -1$ and $x = 1$. Substituting these values back into $K^w l^x m^y g^z$ yields the dimensionless variable $Kl/(mg)$. This new dimensionless variable appears as a governing parameter in the equations of motion for mass-spring models of running, where it has variously been called 'specific spring length' (Blickhan, 1989), 'dimensionless leg stiffness' (McMahon and Cheng, 1990; Bullimore and Burn, 2006), or 'relative stiffness' (Blickhan and Full, 1993). Bullimore and Donelan (2008) have also derived the same dimensionless parameter in a dimensional analysis of running. Here we will refer to $Kl/(mg)$ as 'relative stiffness'.

The need for an elasticity parameter to describe compliant gaits has been widely discussed. Alexander (2003) has suggested using a 'Strouhal number' fl/U with f as the natural frequency of the mass-spring system. Unfortunately, the natural frequency of the system is impractical to measure if it is entrained by forcing, as must be the case in active locomotion. Alexander's suggested alternative of defining f as stride frequency is invalid if the Froude number and relative stride length are also to be considered, because stride frequency is implicit in speed and stride length, and only one physical relationship can connect the variables in a valid dimensional analysis. The so-called 'Groucho number' (McMahon et al., 1987), constructed as the ratio of a Froude number to a Strouhal number, suffers from the same practical limitations. These problems can be avoided by using a formal dimensional analysis to derive the dimensionless variables, as we have done here.

In summary, we have shown that the dimensional relationship describing the dynamics of mass-spring running may be rewritten as a dimensionless one between the Froude number, relative stride length, leg angle, and relative stiffness. This relationship is equivalent to the one derived by Bullimore and Donelan (2008) in their dimensional analysis of running, and we may write the resulting equality constraint as

$$
\frac{U^2}{gl} = \Phi\left(\frac{\lambda}{l}, \frac{Kl}{mg}, \Theta\right)
\tag{3.8}
$$

where Φ is some new and undetermined function. The appearance of mass in Eq. 3.8 shows that, in contrast to stiff gaits, compliant gaits are affected by the mass of the body. Physically, this is because the compression of the leg spring increases as mass is added, which changes the dynamics of the transfer between kinetic and elastic-potential energy.

3.5.2 Empirical constraints upon running

Dimensional analysis is sufficient to predict a set of four dimensionless variables that will be related in running. As was the case for walking, we can go no further in exploring the physical constraints among these without formulating an explicit model of the system. Such a model would not be tractable, however, so we will only explore the answer empirically. Surprisingly, there has been only one empirical study (Donelan and Kram, 2000) in which all four dimensionless variables were measured together.[3] This study involved monitoring ten human subjects of various leg length (l) running at fixed speeds (U) on a treadmill, under conditions of simulated reduced gravity (g). The Froude number ($U^2/(gl)$) was therefore fixed experimentally, so we have plotted how relative stride length varies with Froude number in running humans, overlain upon Alexander and Jayes's classic dataset for 11 species of walking and running mammal (Figure 3.7). It is obvious that the relationships are quite similar, even

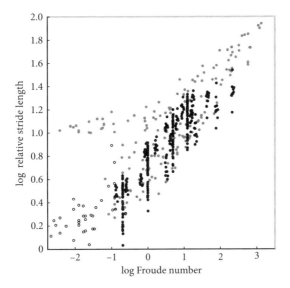

Figure 3.7 Plot of log relative stride length versus log Froude number for 11 species of walking (open circles) and running (grey circles) mammal (Alexander and Jayes, 1983), and for ten human subjects (black circles) running on a treadmill under conditions of simulated reduced gravity (Donelan and Kram, 2000). Original data plotted by kind permission of R. McNeill Alexander and Max Donelan.

[3] Another study of increased gravity running in humans (Cavagna et al., 2005) records relative stiffness but estimated it from a mass-spring model, rather than measuring it directly (G. A. Cavagna, personal communication). A study of running in various other mammals (Farley et al., 1993) recorded Froude number, relative stiffness and angular amplitude, but not relative stride length.

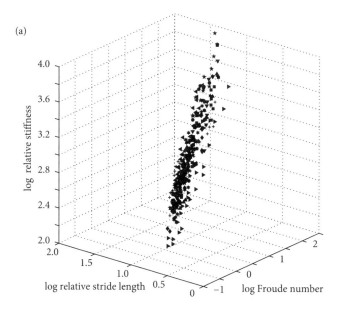

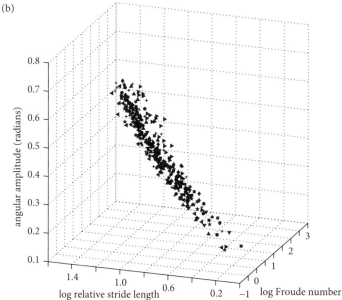

Figure 3.8 Three-dimensional plots of (a) log relative stiffness and (b) angular amplitude plotted against log relative stride length and log Froude number for 10 humans running under conditions of simulated reduced gravity (Donelan and Kram, 2000). Different symbols denote different subjects. It is clear that all four dimensionless variables are closely related, so that the relationship between Froude number and relative stride length is only part of the story. Original data plotted by kind permission of Max Donelan.

in these disparate datasets. Moreover, it is clear from the three-dimensional plots in Figure 3.8 that the four dimensionless variables are indeed closely related, so we may in principle be able to use these data for running humans to shed light in a more general sense on the physical constraints upon running.

Part of the reason why Froude number and relative stride length scale empirically is that both have leg length l in their denominator. This is appropriate if the underlying physical model is known to be correct, because it enforces the necessary constraints upon how the dimensions of the different physical variables can combine. However, it is a nuisance if we are trying to test statistically whether the dimensionless variables are correlated, because having the same measured variable in the denominator can produce spurious correlations between them. As it happens, in Donelan and Kram's study, leg length l was treated as a constant for each of the 10 human subjects, so any covariance that it introduces will be eliminated by controlling for subject in our statistical model. The regression of log relative stride length on log Froude number, controlling for subject, is highly significant ($F_{1,308} = 1239, p < 0.0005$), and takes account of any possible power law relationship between the untransformed variables. In total, the regression explains 80% of the variation in log relative stride length, of which only 1% is explained by subject ($F_{9,308} = 2.36, p = 0.014$). This demonstrates that Froude number and relative stride length are indeed likely to be meaningfully related in running.

Some of the residual variation in log relative stride length is expected to result from our failure to account for the systematic effects of variation in relative stiffness and angular amplitude. We can test this hypothesis statistically by fitting a multiple regression in which we add log relative stiffness and log angular amplitude as new explanatory variables. Log relative stiffness is highly significant when added as a second explanatory variable ($F_{1,307} = 1450, p < 0.0005$), increasing the percentage of the total variation in log relative stride length that is explained by the regression to 97%. Log angular amplitude is also highly significant when added as a third explanatory variable ($F_{1,306} = 47.4, p < 0.0005$), but increases the percentage of the total variation in log relative stride length that is explained by the regression almost negligibly over the preceding model. Because we have taken logarithms, the multiple regression model effectively assumes that the statistical relationship between the untransformed variables is a power-law monomial. We do not know whether this is the correct form for any underlying physical relationship, because the form of that relationship is left undetermined by the dimensional analysis. Nevertheless, the high explanatory power of the model shows that the set of dimensionless variables that we have identified theoretically is sufficient to explain the great majority of the empirical variation in relative stride length in human running.

3.6 Conclusions

We have shown in this chapter how dimensional analysis and explicit analytical modelling can be used to analyse universal physical constraints upon biomechanical systems. Although we have considered brachiation, walking, and running as specific examples, the approach that we have adopted here is completely general, and we use it again to identify physical constraints upon flight performance in Chapter 6. We have chosen to focus upon terrestrial locomotion because idealized models of these systems are well known, and because they have been shown to model the dynamics of real animals rather closely. Physical constraints analogous to those we have identified for terrestrial locomotion will exist for any biomechanical system we might care to examine, however complex it might be. Similar theoretical analyses ought to form the starting point of any empirical exploration of physical constraint in biomechanics.

In general terms, we have considered two distinct types of physical constraint. A constraint that is defined by a mathematical equality is referred to as an 'equality constraint'. The idealized models that we have used to explore brachiation, walking, and running lead to successively more complicated equality constraints, involving two, three, and four dimensionless variables, respectively. Every additional dimensionless variable therefore adds another degree of freedom to the dimensionless design space. Hence, although a real physical system will of course have more degrees of freedom than are contained in any idealized model of it, there may be something to be said for the view that some gaits have intrinsically less scope for variation than others.

The second type of physical constraint that we have considered is defined by a mathematical inequality, and is referred to as an 'inequality constraint'. Inequality constraints set numerical limits upon the values of the variables, and can only be derived with the aid of an explicit theoretical model: they cannot be derived by using dimensional analysis. Inequality constraints bound the design space that organisms are free to explore, and there is some evidence that they are important in enforcing gait changes in terrestrial locomotion. For example, the inequality constraint for inverted pendulum walking contains all of the empirical data points for walking gaits.

Equality constraints and inequality constraints together define the design space that evolution and behaviour are free to explore. Although inequality constraints may be important in understanding gait changes and other discontinuities in behaviour, they will only be important to evolution where organisms press up against the bounds of possibility. Equality constraints, on the other hand, are universal in the sense that any organism with a particular dynamics will always be subject to the same constraint. In the next chapter, we analyse how multivariable constraints of the kind we have discussed here make their effects known in the bivariate scaling relationships that are more often studied in biology. We focus in particular upon how the statistical models that are used to fit scaling relationships differ from the physical models that we have developed in this chapter, paying particular attention to the problems that arise from oversimplifying the analysis of a complex physical system.

4

Scaling

4.1 Introduction

Scaling is a widespread phenomenon in the natural world. It refers to the existence of a power-law relationship of the form $y = bx^a$ between two physical variables x and y, where a and b are constants (Barenblatt, 2003). Scaling laws excite great interest, because they imply that the phenomenon under consideration is self-similar at different scales. For example, the period of a pendulum scales as the square root of its length (see Chapters 1 and 3). Quadrupling the length of a pendulum therefore doubles its period, regardless of its initial length. This is an unusually simple example, because the scaling of pendulum period and length is attributable wholly to physical constraint. In a relationship like this with only one degree of freedom, the effects of natural selection and phylogeny can only possibly manifest themselves in the distribution of data points along the line of constraint. The origins of most bivariate scaling relationships in biomechanics are rarely so simple, however, because the underlying constraints will almost always involve other variables. Consequently, it is reasonable to expect that most empirical scaling relationships in biology will reflect the effects not only of physical constraint, but also of natural selection and phylogeny.

Our aim in this chapter is not to provide a comprehensive review of scaling relationships in biomechanics (see instead McMahon and Bonner, 1983; Schmidt-Nielsen, 1984), but rather to analyse the physics and statistics of scaling relationships. In so doing, we identify some long-standing errors in the way in which biological scaling relationships have been—and continue to be—analysed statistically. We also point to some logical flaws in the inferences that have been drawn from those statistics. Both problems stem from the fact that empirical scaling relationships in biology are, at best, bivariate approximations of multivariable phenomena. This has important consequences for the statistical methods that can be used and for the inferences that we can draw. In particular, it means that we should not treat a bivariate scaling relationship as if it estimated the parameters of an underlying multivariable physical model. In fact, we must be cautious of reading very much at all into the precise numerical value of a fitted scaling exponent.

Unfortunately, an emphasis upon identifying numerical scaling exponents is written into the very language that biologists use to describe scaling relationships, and has been the focus of most empirical studies. We begin this chapter by pointing to the dangers of focussing too much attention on the numerical values of scaling

Evolutionary Biomechanics. Graham Taylor & Adrian Thomas.
© Graham Taylor & Adrian Thomas 2014. Published 2014 by Oxford University Press.

exponents in a model-based discipline like biomechanics. We then analyse the logic of the claim that a bivariate statistical relationship can be used to estimate the parameters of a multivariable physical model, making particular reference to the interspecific scaling of basal metabolic rate with body mass in mammals. Last, we outline the statistical complications that result from fitting an incomplete model, and show why the measurement error models that have been recommended for estimating scaling relationships in biomechanics are almost always inappropriate for this purpose. Statistical misunderstandings run deep through the biological literature on scaling, and our aim in this chapter will be to correct and to clarify some of the advice that has been offered in previous reviews of the subject.

4.2 Some terminological baggage

The analysis of biological scaling relationships originated in biomechanics, with Galileo's famous proposition that the bones of larger animals needed to be thicker relative to the bones of smaller animals in order to sustain the loads to which they were subjected (Galilei, 1638; Figure 4.1). Nevertheless, the terminology of biological

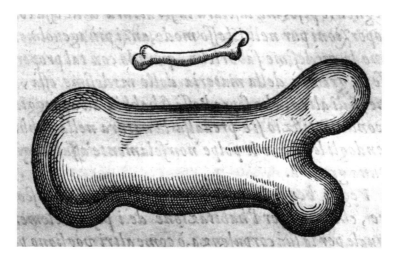

Figure 4.1 Galileo (Galilei, 1638) argued that the bones of larger animals needed to be thicker relative to the bones of smaller animals in order to sustain the loads to which they were subjected. The text that accompanies the figure reproduced here may be translated as follows. *'I once drew the shape of a bone elongated just three times, and swelled in such proportion, that it could do in its own large animal the duty equivalent to that of the lesser bone in the smaller animal; and the shapes are these; wherein you see the disproportioned shape that the enlarged bone becomes.'* This illustration is from the first edition of his text, published in Leiden to escape the reach of the long—and presumably thick—arm of the papal inquisition. Reproduced with thanks to the Principal and Fellows of Jesus College, Oxford.

scaling studies was to crystallize in the field of comparative morphology, where Huxley and Teissier (1936) coined the term 'allometry' to denote the growth of one part of the body at a different rate from that of some standard or of the body as a whole. They used the term 'isometry' to denote identical growth rates. Huxley and Teissier (1936) related these terms explicitly to the exponent of a canonical scaling relationship $y = bx^a$, with isometry implied if $a = 1$, and allometry implied if $a \neq 1$. Nowadays, allometry is understood more generally to refer to systematic changes in shape with size, while isometry is understood to mean that shape does not vary systematically with size (Rayner, 1985). For example, in the scaling of an area (with units of length squared) with a volume (with units of length cubed), isometry would be implied if the scaling exponent was $a = 2/3$. It is standard practice, therefore, to use the numerical exponent of a morphological scaling relationship as a one-dimensional summary of whether, and how, shape changes with size.

It is understandable, but unfortunate, that the same emphasis upon estimating the numerical value of the scaling exponent—often called the 'allometric exponent'—has been carried over universally into biomechanics. Typically, the canonical scaling relationship $y = bx^a$ is linearized by taking logarithms,

$$\log y = \log b + a \log x, \qquad (4.1)$$

and a linear statistical model of one kind or another is then fitted to the data with the aim of estimating the scaling exponent a. This may be reasonable if the scaling exponent is merely being used to summarize the statistical relationship between two variables, as it is in a classical morphometric study. However, there is an almost irresistible tendency in a model-based discipline like biomechanics to attempt to relate the fitted exponent to some underlying physical model. This practice can be highly misleading, because a biomechanical scaling relationship will at best represent a condensed approximation of a more complicated underlying physical function. We are therefore bound to miss important sources of variation if we estimate a relationship between only two variables. As we now show for the example of metabolic scaling, this undermines the whole enterprise of attempting to use a bivariate statistical relationship to identify parameters of an underlying multivariable physical relationship.

4.3 Physical models of scaling relationships

Arguments over the nature of the relationship between metabolic rate (R) and body mass (m) have rumbled on and off for well over a century. Ever since Kleiber first argued that metabolic rate scaled as $R \propto m^{3/4}$ rather than $R \propto m^{2/3}$ in mammals (Kleiber, 1932), the debate has focussed all too often upon establishing and explaining the 'true' value of the scaling exponent across diverse taxa (for recent examples, see White and Seymour, 2003; Savage et al., 2004b). Recent reviews of the subject have been more catholic in their approach, arguing that there may be many different scaling relationships, bounded at one extreme by surface area constraints on the flux of heat, waste, and resources, and bounded at the other extreme by volume constraints on energy use and power production (Glazier, 2005, 2010). Upon this view, metabolic rate is expected to scale as anything between $m^{2/3}$ and m^1 for any set of physiologically similar organisms. Nevertheless, even among those authors who find no evidence for a universal scaling exponent, there remains a universal tendency to attempt to estimate the particular numerical scaling exponents of specific taxa, as if these were in themselves unique (e.g. Duncan et al., 2007; Sieg et al., 2009; Clarke et al., 2010; White, 2011).

The most recent bout of activity in this field was triggered by the publication of a theoretical model predicting that metabolic rate should scale as the three-quarters power of body mass (West et al., 1997). The West-Brown-Enquist model was derived from a model of fluid flow through a hierarchical resource distribution network, but proponents of the theory argue that it can be extended to fractal-like resource distribution networks of all kinds, so as to explain a diverse range of other biological phenomena said to exhibit quarter-power scaling, including mitochondrial densities (West et al., 2002), ontogenetic growth rates (West et al., 2001), allocation of plant production (Enquist and Niklas, 2002), life history timings (Gillooly et al., 2002; Savage et al., 2004a), and even population growth rates (Savage et al., 2004a). These are grand claims, which merit close scrutiny, but our focus here will be upon analysing the physical model of the circulatory system that is the basis of the West-Brown-Enquist model (West et al., 1997). Our specific aim in so doing is to analyse the connection, such as there may be, between complex multivariable physical models and bivariate scaling relationships. Hence, although we do not give detailed consideration to competing theoretical models (e.g. Banavar et al., 1999, 2002; Glazier, 2010), this should not be taken to mean that we disregard them.

4.3.1 *The West-Brown-Enquist model*

The original presentation of the West-Brown-Enquist model (West et al., 1997) has been criticised for its lack of clarity by Kozłowski and Konarzewski (2004) and others. The model has since been derived with greater rigour and under a weaker set of assumptions by Etienne et al. (2006). Neither presentation is easy to follow, however, so we start by providing a simpler derivation of the general result, focussing upon the original formulation of the model as a description of the flow through the mammalian circulatory system. We begin by assuming that the organism's metabolic rate (R) is proportional to the volume rate of flow (Q) at some level of a closed, fluid-filled network used to distribute resources. This is only reasonable if the rate at which resources are transferred to or from the network is limited by bulk flow rate, rather than by their own rate of diffusion.[1] This would not be true of many larger nutrients, which

---

[1] Without this assumption, Eq. 4.2 would be a trivial mathematical identity, rather than a meaningful physical relationship.

tend to be diffusion limited, but it holds under normal circumstances for oxygen uptake in mammals. We may therefore write

$$R = vQ \qquad (4.2)$$

where v measures the amount of resource consumed per unit volume of fluid. For example, if metabolic rate is measured as rate of oxygen consumption, then Eq. 4.2 is just a statement of Fick's principle, where v equates to the difference between the arterial and venous oxygen content of the blood.

We will further assume that the fluid is incompressible and that the network branches hierarchically. The law of conservation of mass now implies that the volume rate of flow is the same at all levels of the network, but it will prove convenient to calculate this at the last level of the network, which in mammals corresponds to the capillaries. Volume rate of flow is defined as the speed of the flow multiplied by the cross-sectional area through which it passes, so we may rewrite Eq. 4.2 as

$$R = v N_c A_c u_c \qquad (4.3)$$

where N_c is the total number of capillaries, A_c is the mean cross-sectional area of a capillary, and u_c is the mean flow speed in a capillary. In the terminology that we introduced in the preceding chapter, this equation represents an equality constraint upon the scaling of metabolic rate in organisms with hierarchically structured resource distribution networks. This constraint does not involve body mass explicitly, but the number of capillaries must obviously increase with body mass; and our next step is to model how it does so.

In order to make further progress, West et al. (1997) make several detailed assumptions about the geometry of the network, which all amount to assuming that one quantity or another is preserved between its different levels. For our present purposes, we will treat these as abstract geometric assumptions of the model, and refer the reader to West et al. (1997) for supporting arguments. First, West et al. assume that the network is space filling, which means that if we conceive of each vessel as being surrounded by a sphere having the same diameter as its length, then the total volume of these spheres would be the same at all levels. Second, they assume that there is a single vessel at the first level of the network and that the number of vessels increases by a constant branching factor at every subsequent level. Third, they assume that the network is area preserving over most of its length, which means that its cross-sectional area is the same between levels. This assumption is relaxed for the last n levels of the network, in which the total cross-sectional area is assumed to increase in such a way that the accompanying decrease in the length of the vessels causes the total volume of fluid to remain the same between levels.

With this geometry, it can be shown (Box 4.1) that the number of capillaries scales as

$$N_c \approx \left[\frac{V \left(1 - v^{-1/3}\right) v^{n/3}}{A_c l_c} \right]^{3/4} \qquad (4.4)$$

Box 4.1 Model of the scaling of a hierarchical resource distribution network.

Here we present a simple geometric derivation of the result given by West et al. (1997) for the scaling of capillary number in the mammalian circulatory system. The total volume of fluid (V) contained between the aorta and the capillaries is

$$V = \sum_{i=0}^{c} N_i A_i l_i$$

where N_i, A_i, and l_i are the number, mean cross-sectional area, and mean length, respectively, of vessels at the ith level of the network. It is assumed that the total cross-sectional area of the network is the same at all levels up to and including the kth level. This implies that $N_i A_i = N_k A_k$ for $i \in \{0, \ldots, k\}$, which entails that the cross-sectional area of these vessels scales inversely with the number of branches. From the kth level up, it is assumed that the total cross-sectional area increases in such a way that each level contains the same volume of fluid, which implies that that $N_k A_k l_k = N_c A_c l_c$. Combining these assumptions, we may make the substitution $N_i A_i = N_c A_c l_c / l_k$ for $i \in \{0, \ldots, k\}$. Assuming that levels above the kth comprise a negligibly small proportion of the total volume, we may write

$$V \approx N_c A_c l_c \sum_{i=0}^{k} \frac{l_i}{l_k}.$$

Next, it is assumed that the network is space filling. This amounts mathematically to assuming that $l_i / l_k = (N_k / N_i)^{1/3}$ at all levels, so we have

$$V \approx N_c A_c l_c \sum_{i=0}^{k} \left(\frac{N_k}{N_i} \right)^{1/3}.$$

Finally, it is assumed that there is one vessel at the zeroth level of the network, and that this number increases by a constant branching factor (ν) at every subsequent level. This implies that $N_i = \nu^i$, and also that $N_k = N_c \nu^{-n}$, where $n = c - k$. Hence,

$$V \approx \frac{N_c^{4/3} A_c l_c}{\nu^{n/3}} \sum_{i=0}^{k} \nu^{-i/3}.$$

The summation term is now just a geometric series with common ratio $\nu^{-1/3}$. This asymptotes rapidly towards $\left(1 - \nu^{-1/3} \right)^{-1}$ as k becomes large, so we may write

$$V \approx \frac{N_c^{4/3} A_c l_c}{\left(1 - \nu^{-1/3} \right) \nu^{n/3}}$$

which we may rearrange to give the number of capillaries as

$$N_c \approx \left[\frac{V \left(1 - \nu^{-1/3} \right) \nu^{n/3}}{A_c l_c} \right]^{3/4}.$$

This is the origin of the three-quarters power scaling prediction in the model of West et al. (1997).

where V is the total volume of fluid contained between the aorta and the capillaries, l_c is the mean length of a capillary, and v is the branching factor relating the number of vessels at one level of the network to the number of vessels at the next. The degree of approximation involved in this equation is acceptable provided that the number of levels in the network is large in its own right, and is large in comparison with the number of levels over which the cross-sectional area of the network is not preserved (n). The approximation is therefore expected to break down for smaller mammals (West et al., 1997). Eq. 4.4 is not a physical constraint, in the sense that there is no physical reason why a hierarchically structured resource distribution network must have the geometry that we have assumed. However, West et al. (1997) have provided theoretical support for the notion that Eq. 4.4 represents a solution to an optimization problem where energy dissipation within the network is minimized.

Eq. 4.4 is the kernel of the model from West et al. (1997) and is the origin of their three-quarters power scaling prediction. It does not involve body mass explicitly, but can be made to do so by noting that the mass of fluid contained between the aorta and the capillaries is necessarily some fraction (η) of total body mass. We may therefore write the identity $V = \eta m/\rho$, where ρ is the density of fluid and m is body mass. Making use of this identity and substituting Eq. 4.4 into Eq. 4.3, we arrive at the following equation for metabolic rate:

$$R \approx v A_c^{1/4} u_c \left[\frac{\eta \left(1 - v^{-1/3}\right) v^{n/3}}{\rho l_c} \right]^{3/4} m^{3/4}. \tag{4.5}$$

This multivariable equation encapsulates the West-Brown-Enquist model of metabolic rate in mammals. No equation like it appears in their original paper (West et al., 1997), because they treated all of the terms except metabolic rate and body mass as scale-invariant quantities. With this assumption, West et al. were able to replace Eq. 4.5 with the simple proportionality $R \propto m^{3/4}$.

4.3.2 Metabolic scaling

The West-Brown-Enquist model was developed to explain the scaling of metabolic rate with body mass. This is a scaling relationship between two variables, but the physical model that has been proposed to explain it contains no fewer than nine! The obvious way forward is not to treat all of the missing variables as invariants as West et al. (1997) and others have done, but rather to make use of their natural variability to determine whether they combine appropriately to predict metabolic rate (Etienne et al., 2006). For example, it would be surprising to find that capillary length and area explained a large proportion of the variation in metabolic rate not already explained by body mass, unless metabolic rate were somehow limited by the properties of the circulatory system. Unfortunately, with the exception of capillary cross-sectional area (Figure 4.3), few of the other terms in the West-Brown-Enquist model have been measured across species. Any progress on this front will need to await new empirical data—an interesting opportunity for future research.

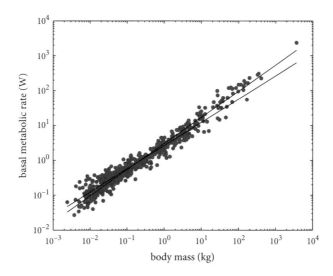

Figure 4.2 Measurements of basal metabolic rate (BMR) versus body mass for 626 species of mammal, from the dataset compiled by Savage et al. (2004b). The two lines have been drawn through the geometric mean of the data to show BMR scaling as either the two-thirds or three-quarters power of body mass.

In the meantime, many have tried to use the supposed ubiquity of quarter-power scaling to confirm the West-Brown-Enquist model (e.g. Savage et al., 2004b), asking whether the 'true' value of the scaling exponent of metabolic rate with body mass is two-thirds or three-quarters (Figure 4.2). Besides erecting a false dichotomy (Glazier, 2005), this question presupposes that there exists a unique underlying scaling relationship between metabolic rate and body mass. We know that this cannot be the case, however, because any unique physical relationship between metabolic rate and body mass is bound to involve other variables. Hence, if any of these other variables is correlated with body mass, then the empirical scaling of metabolic rate with body mass will obviously differ from the naive expectation that $R \propto m^{3/4}$.

For example, capillary cross-sectional area has been found to scale as roughly the one-sixth power of body mass (Dawson, 2001; Figure 4.3). Metabolic rate depends upon the one-quarters power of capillary cross-sectional area in Eq. 4.5, so even this known scale-invariance is sufficient to add one-twenty-fourth to the predicted exponent of the scaling of metabolic rate with body mass in the West-Brown-Enquist model. A difference of one-twenty-fourth may seem trivial, but it is half of the absolute difference between an exponent of two-thirds or three-quarters. It is clear, therefore, that we cannot expect to be able to estimate the exponent of body mass in Eq. 4.5 from the scaling of metabolic rate and body mass if any of the other variables in the equation are correlated with body mass. There is a more subtle problem, however, which arises from the statistics of line fitting. Even if all of the other variables in

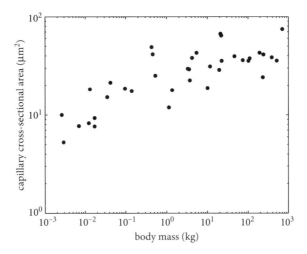

Figure 4.3 Measurements of mean cross-sectional area of a capillary versus body mass for 33 species of mammal, including several different varieties of single domesticated species. Capillary cross-sectional area is calculated from measurements of total capillary volume and surface area given in Table 3 of Gehr et al. (1981), assuming a circular cross-section. By combining data from closely related species, Dawson (2001) has estimated that capillary cross-sectional area scales as approximately the 1/6 power of body mass. In any case, it is evident that capillary cross-sectional area is not scale invariant, contrary to the assumption made by West et al. (1997) in their model of metabolic scaling.

Eq. 4.5 did happen to be uncorrelated with body mass, and even if the West-Brown-Enquist model were an accurate description of reality, we still would not be able to estimate the exponent of body mass in Eq. 4.5 using bivariate statistics. To explain this point fully, we will need to delve into the statistics of line fitting.

4.4 Statistical models of scaling relationships

Physical models of the kind just discussed are rarefied abstractions of the real world, but within their assumptions they are exact and deterministic. Statistical models are neither exact nor deterministic, because they are built to deal with the random variation that appears when we attempt to apply a simplified model to the real world. There are three distinct sources of such random variation. First, any measurement that we make is subject to error, so we do not actually observe the true values of the variables that we are measuring. This is known as measurement error. Second, even the unobserved true values of the variables are unlikely to be perfectly related by any statistical model that we fit, because we will rarely measure all of the variables that are important, and may not be fitting the appropriate form of relationship between them anyway. This is known as equation error. Third, if the unobserved true values of the variables are random samples from some wider population—a species, say—then they too will be subject to random variation themselves. Any statistical model that we use to estimate bivariate scaling relationships in biology must be able to take account of all three sources of random variation. We now outline just such a model, known in the statistical literature as an equation error model (Cheng and Van Ness, 1999). True equation error models are common in econometrics, but have been little used in the biological literature, despite being mentioned from time to time (e.g. Harvey and Pagel, 1991; Riska, 1991; Kelly and Price, 2004; Warton et al., 2006).

4.4.1 Equation error models

We will deal only with linear models here, so if our aim is to fit a scaling relationship of the form in Eq. 4.1 to a given pair of variables, then it is implicit that we will have taken logarithms of both variables prior to analysis (see Eq. 4.1). We discuss the consequences of making this transformation at the end of this section. Because our measurements are made with error, we do not observe the true values of these variables directly, but instead measure

$$X_i = x_i + \delta_i, \quad Y_i = y_i + \epsilon_i, \quad i = 1, \ldots, n \tag{4.6}$$

where X_i and Y_i are the measured value of each variable for the ith measurement, and where x_i and y_i are the true underlying values of the variables. The measurement errors δ_i and ϵ_i are random variates (i.e. values drawn at random from some specific distribution), as are the measurements X_i and Y_i. In practice, we do not expect even the true values of the variables x_i and y_i to be perfectly linearly related, so the model that we fit is hypothesized to be of the form

$$y_i = \alpha + \beta x_i + q_i, \quad i = 1, \ldots, n \tag{4.7}$$

where α and β are parameters that we will estimate, and where the equation error q_i is a random variate that is assumed to be independent of x_i for all i. The equation error q represents the extent to which the relationship between the true values of the variables x and y deviates from linearity. The combination of Eqs. 4.6 and 4.7 is known as an equation error model.

We will need to make some further assumptions in order to estimate the slope and intercept of Eq. 4.7. We will assume that the measurement errors (δ_i, ϵ_i) and the equation errors (q_i), are independent, and will also assume that they are normally distributed with mean zero and variance σ_δ^2, σ_ϵ^2, and σ_q^2. In reality, the equation errors q_i are unlikely to be independent in a cross-species study, because related species are likely to share variation in important but omitted variables. We ignore this complication for the time being, but discuss in Chapter 5 how comparative data can be transformed to meet this assumption. Even with these assumptions, we can only estimate the slope and intercept of the relationship in Eq. 4.7 if we have prior knowledge of the measurement error variance σ_δ^2 in X. If this is known, then we may estimate the slope of the relationship in Eq. 4.7 as

$$\hat{\beta} = \frac{s_{XY}}{s_{XX} - \sigma_\delta^2} \tag{4.8}$$

where s_{XX} is the sample variance of X, and where s_{XY} is the sample covariance of X and Y (Cheng and Van Ness, 1999). Eq. 4.8 is a consistent estimator of the slope of the relationship in Eq. 4.7, in the sense that it converges in probability to the true value of the parameter as the sample size increases to infinity. This is the least that we should require of any reasonable estimator, although it is not true of some of the other estimators that are used in the literature (see below).

The inclusion of equation error in Eq. 4.7 introduces a fundamental asymmetry in the treatment of the x and y variables, because the equation errors (q_i) combine additively with the deterministic part of the relationship ($\alpha + \beta x_i$) to give y_i. Consequently, we will fit a fundamentally different relationship depending upon which variable we call x and which we call y. This asymmetry disappears only in the unlikely case that there is no equation error (i.e. if the underlying relationship between the true values of the variables x and y is truly linear). Assigning the equation error to one or other variable is necessary in order to make the model identifiable (i.e. in order to ensure that there is a unique solution for the parameters α and β). However, it is arbitrary which variable we call x and which we call y. If we switch their identity, then it can be shown using the Cauchy-Schwarz inequality that the slope of the resulting relationship will be steeper than the first. This ambiguity is appropriate, because in fitting a model with equation error, we are admitting that our model does not correctly capture the form of the underlying relationship between the variables.

Since it is arbitrary how we choose to assign the equation error, there is no reason to expect that we should be able to fit only one line to the data. For example, should we choose to regard the shrews (Soricidae) in Figure 4.4 as having unusually high metabolic rate given their body mass (thereby assigning the error to metabolic rate), or unusually low body mass given their metabolic rate (thereby assigning the error to body mass)? These are simply different ways of looking at the same data (Warton et al., 2006). In fact, the two lines that can be fitted using an equation error model really represent two extremes of a continuum, according to whether we choose to assign all of the equation error to one variable or the other. This ambiguity does not represent a deficiency in the modelling framework: rather, it reflects the fact that we cannot uniquely identify the relationship between two imperfectly related variables, without making additional assumptions that entail an asymmetry in the treatment of the variables. This ambiguity may be hard to accept if we had hoped to estimate a single underlying relationship between the variables, but we deceive ourselves if we do otherwise.

4.4.2 Regression models

The asymmetry of the treatment of the variables in an equation error model parallels the well-known asymmetry of the x and y variables in a regression model. In fact,

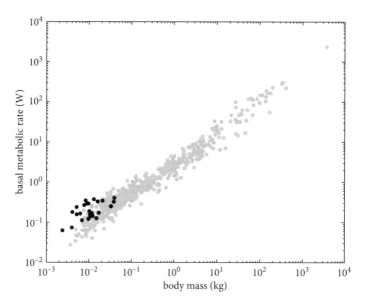

Figure 4.4 Measurements of BMR versus body mass for 626 species of mammal, from the dataset compiled by Savage et al. (2004b). The points in black denote members of the shrew family (Soricidae), and will evidently not be modelled well by any relationship fitted to the data as a whole. However, it is arbitrary whether we choose to assign the resulting equation error to metabolic rate or to body mass. The former amounts to saying that shrews have an unusually high metabolic rate given their body mass; the latter amounts to saying that shrews have an unusually low body mass given their metabolic rate.

a regression is just a special case of an equation error model, where the measurement error variance (σ_δ^2) in X is zero. Substituting $\sigma_\delta^2 = 0$ into Eqs. 4.6 and 4.7 and rearranging yields the classical regression model:

$$Y_i = \alpha + \beta x_i + e_i, \quad i = 1, \ldots, n \tag{4.9}$$

where $e_i = (q_i - \epsilon_i)$ is the combined measurement error and equation error. Because the measurement error (ϵ_i) and the equation error (q_i) are combined in a regression model, equation error is not discussed separately in the classical regression literature (Cheng and Van Ness, 1999). Substituting $\sigma_\delta^2 = 0$ into Eq. 4.8 yields

$$\hat{\beta} = \frac{s_{XY}}{s_{XX}}, \tag{4.10}$$

which is the usual regression estimator.

Because the measurement error variance (σ_δ^2) in X cannot be negative, it is obvious by inspection of Eqs. 4.8 and 4.10 that the regression slope estimated by Eq. 4.10 cannot be steeper than the slope estimated by Eq. 4.8 for the equivalent equation error model (Figure 4.5). Failing to account for measurement error in X therefore

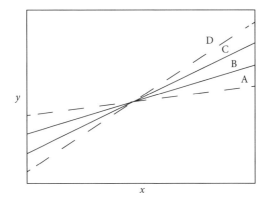

Figure 4.5 Two linear relationships can be fitted using an equation error model (solid lines), depending upon whether the equation error is assigned to the variable plotted on the *y*-axis (B) or the *x*-axis (C). Likewise, two linear relationships can be fitted using a regression model (dashed lines), depending upon whether all of the error is assigned to the variable plotted on the *y*-axis (A) or the *x*-axis (D). Provided that the measurement errors in the two variables are uncorrelated, the qualitative arrangement of the lines will always be as shown in the figure. Only if the data are perfectly correlated will the four lines coincide.

diminishes our estimate of the slope. This effect is known as attenuation, and means that the slope of the regression in Eq. 4.9 cannot be steeper than the slope of the line calculated using the equation error model in Eqs. 4.6 and 4.7. The equation error model can therefore be regarded as a modified regression model, corrected for attenuation (Carroll and Ruppert, 1996). Hence, if we switch the identity of the variables, then it is easy to see that the two lines that we could estimate using an equation error model must lie between the two lines that we could estimate using a regression model (Figure 4.5). This is an important property of the equation error model, but holds only if the errors are uncorrelated, as we have assumed above.

In cases where the measurement error in one of the variables is small, it may be reasonable—as an approximation—to fit a scaling relationship by using the regression estimator (Eq. 4.10) instead of the equation error model estimator (Eq. 4.8). For example, because it is usually possible to measure body mass with a high degree of accuracy, it is perfectly reasonable to regress log BMR on log body mass (Figure 4.6). In fact, this is the standard practice in metabolic scaling studies (Glazier, 2005). For reasons that we have just discussed, we are guaranteed not to overestimate the slope of the relationship if we regress log BMR on log body mass. Hence, if our aim is to test whether the slope of the relationship is greater than two-thirds, then we will be making a conservative hypothesis test if we use the regression of log BMR on log body mass. As it happens, regressions of log BMR on log body mass across mammals often do find a regression slope that is significantly greater than two-thirds, both before and after correcting for phylogenetic error covariance (e.g. Duncan et al., 2007;

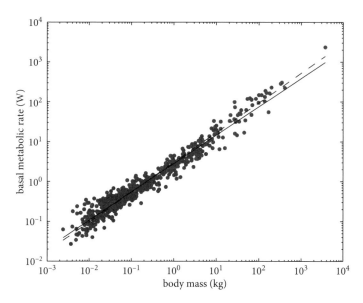

Figure 4.6 Measurements of BMR versus body mass for 626 species of mammal, from the dataset compiled by Savage et al. (2004b). The solid line plots the regression of log BMR on log body mass. The dashed line plots the regression of log body mass on log BMR. We do not have available the information on measurement error variance that is necessary to identify an equation error model, but any line that we could fit using such a model would have to lie between these two regression lines, which happen in this case to be close anyway.

Clarke et al., 2010). The conclusion that the slope of the relationship is greater than two-thirds therefore seems to be robust.

On the other hand, the regression of log BMR on log body mass (Figure 4.6) is not the only reasonable estimate of the relationship between the variables. If we could quantify the measurement error variance in log BMR, then it would be equally reasonable to fit an equation error model with log BMR as its *x* variable. This relationship would have a steeper slope than the regression of log BMR on log body mass (Figure 4.5), but is an equally valid estimate of the relationship. However, we know that the slope of this alternative equation error model cannot be steeper than the slope of the regression of log body mass on log BMR. Hence, as the two regression lines happen to be very close to each other in this case, it will not make much difference to our estimate of the relationship if we choose to switch the *x* and *y* variables in our equation error model (Figure 4.6).

4.4.3 Errors-in-variables models

At this point, readers who are familiar with the biological scaling literature may be wondering why we have not yet mentioned errors-in-variables models, such as the major axis or reduced major axis models. An errors-in-variables model can be thought of as a degenerate case of an equation error model where the equation error q_i is identically zero. This makes them inappropriate in cases where equation error cannot be ignored, which is the case for most bivariate relationships in biology. Nevertheless, because errors-in-variables models are so widely misused (Carroll and Ruppert, 1996), it is important that we discuss them here. Substituting $q_i = 0$ into Eqs. 4.6 and 4.7 yields the errors-in-variables model

$$X_i = x_i + \delta_i, \quad Y_i = y_i + \epsilon_i, \quad y_i = \alpha + \beta x_i, \quad i = 1, \dots, n. \tag{4.11}$$

If we know the measurement error variance (σ_δ^2) in X, then we can obviously just use the estimator that was already given for the equation error model in Eq. 4.8. However, this entails asymmetric treatment of the variables, and an important feature of errors-in-variables models is that they permit the variables to be treated symmetrically.

For example, if the ratio of the measurement error variances $\lambda = \sigma_\epsilon^2/\sigma_\delta^2$ is known, then a consistent estimator of the slope of the relationship is

$$\hat{\beta} = \frac{s_{YY} - \lambda s_{XX} + \sqrt{\left(s_{YY} - \lambda s_{XX}\right)^2 + 4\lambda s_{XY}^2}}{2s_{XY}} \tag{4.12}$$

where s_{YY} is the sample variance of Y. This is widely referred to as the general structural relation in the biological scaling literature (Rayner, 1985), but the use of this terminology ignores the careful distinction that statisticians make between a structural and a functional relationship (see below). In the special case that $\lambda = 1$, Eq. 4.12 defines what is then called the major axis of the data. However, it is essential to note that knowledge of λ can only allow us to identify the true underlying relationship between the true variables in the unlikely event that they are perfectly linearly related, so the estimator in Eq. 4.12 is of little use in most practical situations in biology.

In practice, we will often have no prior knowledge about the measurement error variances. In this situation, Rayner (1985) has advocated using the reduced major axis estimator, or geometric mean functional relationship:

$$\hat{\beta}_4 = \text{sign}\left(s_{XY}\right) \sqrt{\frac{s_{YY}}{s_{XX}}} \tag{4.13}$$

on the grounds that this is the maximum likelihood estimate of the slope of the relationship if nothing is known about the errors. This is incorrect, although the reason is quite technical and depends specifically upon whether the values of the x_i are treated as unknown constants (the so-called functional case) or as random variates (the so-called structural case). Whilst it is true that Eq. 4.13 is a solution of the likelihood equations in the functional case, it has long been known that the likelihood equations are unbounded and that the reduced major axis identifies a saddle point rather than a maximum in the likelihood function (Solari, 1969). In the structural case, the model is not identifiable at all if the measurement error variances are both unknown (at least, not without further side assumptions), so the reduced major axis estimator does not even arise in this case (Cheng and Van Ness, 1999). In fact, it can be shown that the reduced major axis does not provide a consistent estimate of the slope of the relationship in Eq. 4.11, even in the absence of equation error (Sprent and Dolby, 1980; Cheng and Van Ness, 1999). It is therefore regrettable that the reduced major axis estimator has been so widely used to estimate scaling relationships in biomechanics (e.g. Rayner, 1985; Alerstam et al., 2007). The obvious appeal of the reduced major axis estimator is that it requires no knowledge of the measurement error variances; unfortunately this is a siren song.

4.4.4 Logarithmic transformation

We have focussed most of our attention so far upon how to deal appropriately with the different kinds of error that are associated with scaling relationships in biology. We now look briefly at how logarithmic transformation affects both the error structure of the model and the fit of the model to the untransformed data. Logarithmic transformation is necessary if we are to analyse a scaling relationship of the form $y = bx^a$ using a linear statistical model. However, as non-linear statistical models—including non-linear equation error models—are now widely available, linearization may not be sufficient justification for transforming the data (Packard et al., 2011). A stronger justification is that most statistical models assume that the errors are identically distributed, which tends not to be the case when measurements are ranged over several orders of magnitude. Larger measurements tend to be associated with larger measurement errors, and it may therefore be more reasonable to assume a multiplicative rather than additive error model. This should always be checked against the data, rather than being assumed automatically (Packard et al., 2011), but in this situation it would usually be indefensible not to take logarithms. It is easy to show by back-transformation that assuming an additive error model after taking logarithms, as we have done above, amounts to assuming a multiplicative error model on the original measurement scale. This is certainly reasonable for the example of metabolic scaling in mammals that we have considered throughout this chapter, for which the available data span nearly seven orders of magnitude in body mass.

Transforming the data in this way also reduces the leverage of measurements made at the larger end of the scale, which is important in the sense that we would not want the estimate of the 'mouse to elephant curve' to be driven wholly by the elephant. However, this means that the fit of the model may be rather poor for these data points, when assessed on the original measurement scale. To illustrate this point, Figure 4.7 plots the regression of log BMR on log body mass after back-transformation. It is more than a little disconcerting to see how poorly the fitted relationship fits the data points for larger mammals, when viewed on the original measurement scale (see also White, 2011). Although the fitted scaling relationship looks to be performing well across seven orders of magnitude of variation in body mass when plotted on a logarithmic scale, plotting the relationship on an arithmetic scale gives an altogether different impression (Figure 4.7). We should not lose sight of the fact that the fitted relationship describes the distribution of observations in the logarithmic domain, and not the arithmetic one (Packard et al., 2011). Hence, it is always an important

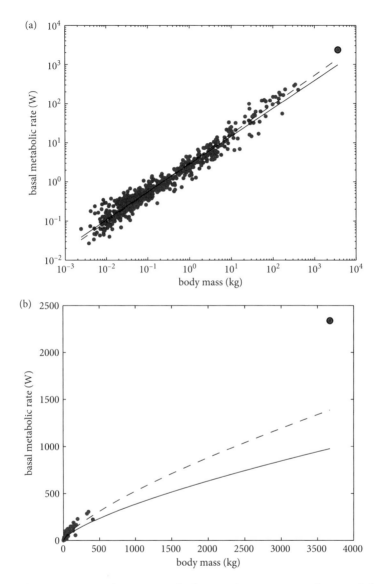

Figure 4.7 Measurements of BMR versus body mass for 626 species of mammal, from the dataset compiled by Savage et al. (2004b). The solid lines plot the regression of log BMR on log body mass. The dashed lines plot the regression of log body mass on log BMR. In (a), the data are plotted in the logarithmic domain in which the relationships were fitted. In (b), the data are plotted on their original measurement scale, and the regressions have been back-transformed accordingly. It is clear that the circled data point for the elephant is not remotely well explained by the fitted relationship when viewed on the original measurement scale (b), even though it is not an obvious outlier when plotted in the logarithmic domain (a).

reality check to back-transform the data before drawing bold conclusions about the universality of a scaling relationship fitted to log-transformed data.

4.5 Conclusions

Scaling relationships in biomechanics are, at best, bivariate summaries of multivariable phenomena. For example, the West-Brown-Enquist model (West et al., 1997) was developed to explain the scaling of metabolic rate with body mass in mammals, but the model itself involves no fewer than nine different variables. Much of the scatter that exists about these relationships is therefore likely to be attributable to important variables that are missing from the fitted equation. It follows that the scatter around a scaling relationship can sometimes be more interesting than the scaling relationship itself—a point which we take up further in the next chapter. The major axis, reduced major axis, and general structural relation models that have been widely advocated for fitting scaling relationships in biomechanics purport to estimate a unique relationship, but fail to take account of equation error. In consequence, none of these methods is able to provide a consistent estimate of the statistical relationship between variables in the presence of equation error. Furthermore, the reduced major axis estimator is an inconsistent estimator of relationship, even in the absence of equation error. In practice, this makes errors-in-variables models unsuitable for estimating scaling relationships in biology.

Equation error models offer consistent estimates of the statistical relationship between two imperfectly related variables, but they assign all of the equation error to one of the variables and require knowledge of the measurement error variance in the other variable. This asymmetry in their treatment of the variables means that a different relationship will be estimated if we switch the identity of the variables. The aim of identifying a unique underlying relationship is therefore doomed if the fitted model is missing important variables. Underpinning this statistical indeterminacy is a logical indeterminacy. For example, if the underlying physical relationship that connects metabolic rate and body mass involves other variables, as it surely does, then there is no bivariate scaling relationship that uniquely connects metabolic rate and body mass. It is therefore nonsensical to try to establish the 'true' value of their scaling exponent, as if such a thing existed. Asking whether the 'true' value of the scaling exponent of metabolic rate with body mass is two-thirds or three-quarters is like asking whether the king of France is wise: the very question presupposes something that does not exist (see Strawson, 1950). France has no monarch; neither is metabolic rate dictated solely by body mass.

In practice, we will often lack the information on measurement error variance that would be needed to fit an equation error model. Fortunately, a regression model can always be used to estimate a relationship in place of a full equation error model if the measurement error in one of the variables is negligibly small. However, by fitting the regression we are—in effect—choosing to assign all of the equation error to the variable that was measured with error. This is an arbitrary choice, and it is important to recognize that there is an alternative equation error model that could be fitted if we were to switch the identity of the variables. Hence, even though the measurement error may be small in only one of the variables, and even though it will then be obvious which variable to regress on which, we still will not have identified a unique relationship between the variables. In the event that we know nothing at all of the measurement error variances, the very best that we can do is to fit the regression of y on x and the regression of x on y, and to assert that any reasonable estimate of their relationship must fall between the two regression lines (Figure 4.5).

In conclusion, we cannot expect to understand the origin of bivariate scaling relationships, or the scatter about them, without moving to levels of explanation that are consistent with the complexity of the problems in hand. There are two approaches that we might take for this. One is to use a theoretical model to predict what further physical variables need to be considered in our statistical model of the phenomenon, and use these to explain as much of the equation error as possible. This is broadly the approach that we took in Chapter 3, where we used dimensional analysis to identify the smallest set of physical variables that could be used to describe running gaits, and then built up our statistical model accordingly. The second is to abandon any attempt at modelling the detailed physics, and to explain away the equation error using phylogenetic or ecological variation as a proxy for the unmeasured, and possibly unknown, physical variables. This is the approach that we adopt in Chapters 5 and 7, where we explore how the phylogenetically controlled comparative method can be used to analyse the origin of biomechanical scaling relationships and the scatter that exists about them.

5

Phylogeny

5.1 Introduction

In the previous two chapters, we have been most interested in explaining physical variables such as the Froude number or metabolic rate in terms of their physical relationships to other physical variables. We have paid less attention, so far, to the functional (i.e. adaptive) and historical (i.e. phylogenetic) explanations of traits. Adaptation and phylogeny can interact with the physical constraints upon biomechanical systems in either of two ways. First, phylogeny and adaptation may gate physical constraints, by determining the broad class of dynamics that it is possible or beneficial for a particular organism to adopt. For example, there are obvious phylogenetic reasons why birds fly on migration whereas wildebeest do not, but there are presumably also adaptive reasons why birds choose to fly rather than to walk when covering large distances. Second—and much more interestingly—the combined influences of adaptation and phylogeny may themselves be shaped by physical constraints, which determine, for a given class of dynamics, the design space that evolution is free to explore.

In this chapter, we use the scaling of wing area in birds to explore how phylogeny manifests itself in comparative biomechanics. In the context of explaining the empirical relationships between variables, what used to be referred to as the 'effects' of phylogeny are really nothing more than equation error resulting from the presence of shared variation in important but omitted variables. This shared variation is a statistical nuisance, because it results in error covariance along phylogenetic lines. If left unaccounted, this will bias our estimates of the standard error of any parameters we estimate, leading us to overestimate the statistical significance of any results we obtain. This is the primary motivation for the phylogenetically controlled comparative method, which we discuss at some length here, first because it has still scarcely penetrated the biomechanics literature, and second because the precision with which biomechanical problems can be defined yields useful insight into the comparative method itself.

We begin this chapter by introducing a very large comparative dataset on the flight morphology of birds, which we use to explore evolutionary questions about biomechanics throughout the rest of this book. We then use these data to illustrate some important features of the comparative method itself. We start by considering how

Evolutionary Biomechanics. Graham Taylor & Adrian Thomas.
© Graham Taylor & Adrian Thomas 2014. Published 2014 by Oxford University Press.

phylogeny can manifest itself when fitting a relationship between measured variables, before using this to frame our discussion of the theory underpinning the comparative method. Finally, we use the scaling of wing area with body mass in birds as a simple example to illustrate the use of the comparative method in practice.

5.2 Comparative data and their pitfalls

The 10,000 or so species of living birds display substantial variation in their flight morphology. This morphological variation has been accomplished with little modification of the basic flight anatomy, and has become a classic study piece in comparative biomechanics. It is rather surprising, therefore, that no comprehensive survey of avian flight morphology to date has made use of the phylogenetically controlled comparative method. With this in mind, we have assembled a very large dataset containing measurements of body mass, wingspan and wing area for 450 species of bird, gathered from an exhaustive literature search.[1] This is a very large dataset by any standards, and especially by the standards of the biomechanical literature, where the technical difficulty of data collection means that comparative statements must often be made by comparing a very limited number of species. As well as serving as the primary dataset that we use to analyse evolutionary questions through the rest of this book, these data serve to illustrate some of the practical difficulties that arise with comparative datasets—particularly when these are compiled from a variety of published sources.

One general issue with compiling data from different sources is that the resulting sample cannot usually be argued to be a random sample of any wider population. For example, Western Palaearctic species are over-represented in our sample. They account for two-thirds of our data, but less than one-tenth of global bird diversity. Conversely, passerines (Passeriformes) are under-represented in our sample. They account for one-quarter of our data, but one-half of all living species. Indeed, the suboscine passerines (Tyranni), which comprise approximately one-tenth of the diversity of living birds, are represented by only one species in our sample. The most important practical consequence of this is that we will not usually be able to treat the true values of the variables in the models that we fit as random variates, because we cannot argue that our data represent a random sample of any wider population. Among other things, this means that the conclusions that we draw are to be

---

[1] The published data which met the quality control criteria outlined in the text were drawn from: Alerstam et al. (2007); Bruderer and Boldt (2001); Bruderer et al. (2010); DeJong (1983); Greenwalt (1962); Hedenström and Møller (1992); Hertel and Ballance (1999); Kerlinger (1989); Magnan (1922); McGahan (1973); Norberg (1986); Pennycuick (1971, 1990, 1996); Spear and Ainley (1997); Tobalske (1996, 1999); Warham (1977). We also included a small number of unpublished measurements collected by one of us (ALRT) with Ian Owens.

understood as applying specifically to the subset of birds that is represented in our dataset.

When compiling comparative data from different sources, care must be taken to ensure that similar quantities are being compared across studies. The literature on avian flight morphology is especially problematic in this respect, because wing area can be defined in several ways. From an aerodynamic perspective, the appropriate measurement sums the total area of both outstretched wings and the projected area of the body between them (Pennycuick, 1999). Unfortunately, some previous analyses have unwittingly treated measurements made using different definitions of wing area as if they were measurements of the same thing. In the worst case, this can introduce systematic bias to the fitted relationship. For example, in the comparative dataset assembled by Greenwalt (1975) and later reused by Rayner (1988) and Norberg (1990), most of the measurements of wing area for hummingbirds (Trochilidae) exclude the area of body between the wings. This presumably explains why these authors found that wing area scaled oddly with body mass in this particular family of birds. Similarly, what would otherwise have been an exceedingly useful dataset by Mendelssohn et al. (1989), giving body mass, wingspan and wing area for 66 species of African raptor, unfortunately measured wing area excluding the area of the body between the wings. Hence, we have taken care to include in our dataset only those measurements[2] that appear to satisfy the definition of wing area given by Pennycuick (1999).

In addition to morphological data, we also require a phylogeny identifying the interrelationships between the species included in our dataset. Molecular phylogenies are now widely available, but we must still contend with the conflicts that exist between published phylogenies, and with the need to stitch together different phylogenies in order to arrive at a fully resolved tree. The phylogeny of birds is quite well known, but although recent phylogenies are congruent in many respects, several of the higher level relationships remain controversial (Mayr, 2011). We constructed an almost fully resolved phylogeny for the species in our sample, using the phylogenomic study of 19 nuclear loci by Hackett et al. (2008) to establish the higher level phylogeny. We used a miscellany of other molecular phylogenies to resolve the intra-familial relationships. This left us with only 11 polytomies—all of them between congeners, except in the case of rails (Rallidae), for which the intra-familial relationships remain unresolved. We did not use molecular distance data to assign branch lengths to the phylogeny, partly because it would have been impractical to do so, and partly because there is no particular reason to expect phylogenetic error covariance to depend directly upon molecular distance. We discuss this point at greater length below.

---

[2] Where data were drawn from Greenwalt (1962), wing area was back-calculated from aspect ratio and span, as in Alerstam et al. (2007).

5.3 On the origin of phylogenetic non-independence

One of the most important determinants of flight performance is the ratio of body weight to wing area, known as 'wing loading' in the aeronautical and biomechanical literatures. Under the assumption of isometry (Chapter 4), wing loading is expected to scale as the one-third power of body mass. This means that larger birds will have to sustain a greater aerodynamic load per unit wing area, unless flight morphology varies systematically with size to compensate. We can test this by using the kinds of statistical techniques that we discussed in Chapter 4, but in doing so we will need to take due account of the phylogenetic non-independence of the data.

5.3.1 Equation error in comparative data

Phylogenetic non-independence is clearly apparent in the plot of wing area against body mass in Figure 5.1, which compares members of the auk family (Alcidae) with the other birds in our dataset. It is clear that the different species of auk all have unusually small wing area, and hence unusually high wing loading, given their body mass. Were we to add a regression line to the plot, the residuals associated with these 17 data points would all share the same, negative sign. A sign test is sufficient to show that these errors are unlikely to be independent ($p < 0.00001$), so

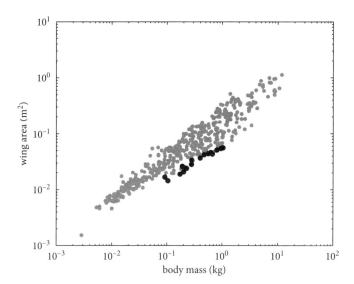

Figure 5.1 Measurements of wing area versus body mass for $n = 450$ species of bird. The points in black denote 17 members of the auk family (Alcidae), which all have a relatively low wing area given their body mass. This will manifest itself as phylogenetic error covariance if we attempt to fit a relationship between wing area and body mass without controlling for phylogeny.

this is a clear example of phylogenetic error covariance. We discuss statistical techniques controlling for phylogenetic error covariance in the next section, but for now we focus our attention upon elucidating its precise origins.

The phylogenetic error covariance in the regression of log wing area on log body mass is attributable to local but systematic variation in wing area, over and above the general systematic scaling of wing area with body mass. Assuming that our measurement technique is not biased along phylogenetic lines, this can only possibly arise as a result of systematic variation in shape or density that is omitted from the deterministic part of the fitted relationship. Hence, in the terminology of the previous chapter, the phylogenetic error covariance is attributable to equation error, resulting from systematic variation in important but omitted variables. In general, if we are certain that our measurement technique is not biased along phylogenetic lines, then phylogenetic error covariance must always be attributable to equation error.

One important consequence of this is that the apparent extent of phylogenetic error covariance is contingent upon the relationship that we fit. Indeed, if we could fit a relationship whose deterministic part explained all of the systematic variation in the dependent variable, then any phylogenetic non-independence would disappear altogether. For example, if we plot wing area against wingspan (Figure 5.2), instead of against body mass, then the auks no longer show such obvious phylogenetic non-independence. Moreover, because wing area is equal to wingspan times mean wing chord (defined as the mean distance from front to back of the wing), a multiple

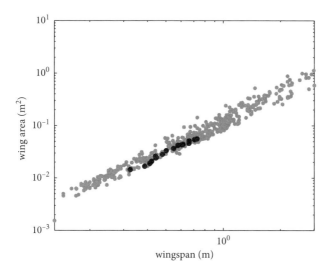

Figure 5.2 Measurements of wing area versus wingspan for $n = 450$ species of bird. The points in black denote 17 members of the auk family (Alcidae). Whereas the auks are obvious outliers in the plot of wing area against body mass (Figure 5.1), they are not such obvious outliers in the plot of wing area against wingspan. The apparent extent of phylogenetic non-independence will vary according to which relationship we consider.

regression of log wing area on log wingspan and log wing chord should have no equation error, and hence no apparent influence of phylogeny: the laws of geometry are oblivious to the evolutionary history of a species. Phylogenetic error covariance is an artefact of fitting an incomplete relationship, and will vanish if we ever manage to fit a complete one.

It follows that what appears to be an influence of phylogeny in one relationship may appear to be the result of adaptation in another. For example, birds that routinely swim underwater tend to have comparatively small wing areas in relation to their body mass. Were we to control statistically for underwater swimming behaviour when regressing log wing area on log body mass, then we would find that the apparent influence of phylogeny on wing area would be weakened, and attributed instead to adaptation for underwater swimming. This is entirely consistent with the widely held view that the phylogenetic non-independence of comparative data reflects niche conservatism among related species (e.g. Grafen, 1989).

5.3.2 Evolutionary models and equation error

The view that phylogeny manifests itself as equation error, at least in the context of explaining the empirical relationships between variables, was first articulated by Grafen (1989). However, it seems to have been lost sight of in the more recent literature, which has followed Felsenstein (1985) in explaining and interpreting phylogenetic error covariance as the outcome of some explicit model of trait evolution (e.g. Freckleton, 2009; Freckleton et al., 2011). The classical line of argument after Felsenstein (1985) begins by asserting that traits evolve through time under a Brownian model of possibly correlated evolution. The key result of this model is that characters are expected to be distributed according to a multivariate normal distribution, with mean equal to the state of the character at some initial time t_0, and variance proportional to the time elapsed since t_0. The Brownian model is sometimes described as a neutral model of evolution (e.g. Butler and King, 2004), but this is misleading: first because selection in response to a randomly fluctuating environment may result in apparently stochastic evolutionary change; and second because correlated evolution under a Brownian model might sometimes be the result of selection, rather than chance or constraint.

More recently, Felsenstein's argument (Felsenstein, 1985) has been extended by using a modified form of Brownian motion described by an Ornstein-Uhlenbeck model to allow for the possibility that traits have a tendency to be attracted towards a particular mean (e.g. Hansen and Martins, 1996). This is intended to account for the action of stabilizing selection, with the attractor perhaps representing some evolutionary optimum. Although the Brownian and Ornstein-Uhlenbeck models differ in important respects, they both predict that the characters will be multivariate normal in their distribution. In practice, however, there will often be no reason to expect or assume that a set of characters should be multivariate normal. On the contrary, we may want or need to treat an explanatory variable as categorical rather than continuous, or as fixed rather than random. Neither of these treatments is consistent with simultaneously assuming that the characters are multivariate normal.

Fortunately, the explanatory variables can be treated in any of these ways by the method of generalized least squares. This method does not require any detailed distributional assumptions to be made about the explanatory variables. Instead, it makes assumptions about the residual error, which is typically assumed to be normally distributed, although not independently so. The assumption of normally distributed error is quite reasonable if the error is equation error resulting from the additive effects of a multitude of omitted but effectively independent random variables. In this case, we expect the error to approach a normal distribution on statistical grounds, regardless of how the omitted variables are distributed, as a result of the central limit theorem.[3] This provides a more general justification for assuming normality of error than the usual assumption of a Gaussian model of trait evolution.

Hence, if generalized least squares models are to be thought of as making assumptions about the evolution of anything, then it is about the evolution of the equation error that results from omitting important variables from the model. This is obviously a rather esoteric concept, because whereas the omitted variables will often be biological traits whose evolution can be described by some meaningful evolutionary model, this is not the case for the equation error. Recall that the equation error is simply that part of the variation in the dependent variable which is left unexplained by the deterministic part of the fitted model (see Chapter 4). It therefore depends not only upon what explanatory variables have been included in the model, but also upon the form of the underlying and fitted relationships. Equation error is not a trait as such, so we should not expect it to be easily related to any explicit model of trait evolution. We return to this important point again later, after discussing how to control statistically for phylogenetic error covariance.

5.4 The comparative method in theory

Felsenstein's method of independent contrasts (Felsenstein, 1985) holds a special place in the literature on the comparative method. For one thing, it remains widely used among practitioners. This is due in part to the popularity of certain software implementations of the method, but also reflects its intuitive evolutionary rationale. In fact, although it can be regarded as just a special case of generalized least squares (e.g. Rohlf, 2001), Felsenstein's method still looms large in most explanations of the comparative method, because of the ease with which it can be explained. In contrast, although generalized least squares is the more general of the two methods, it cannot be easily explained without some knowledge of linear algebra. Both methods can be thought of as involving a linear transformation of the original data, the purpose of

---

[3] If the omitted variables combine multiplicatively, then the equation error is expected to approach a normal distribution after log-transformation of the variables.

which is to cause the transformed data to conform to the usual assumptions of ordinary least squares. We therefore use Felsenstein's method to explain the principle of transforming the data, before giving an informal explanation of the matrix transformation that is involved in generalized least squares. The mathematical details of this transformation are provided in Box 5.1.

5.4.1 Independent contrasts

Although the actual values of a character are not expected to be independent when measured on related species, differences between the measured values are expected to be independent if calculated between sister species on a bifurcating phylogeny. This is because the changes that are associated with each radiation may be assumed to represent independent evolutionary events (Ridley, 1983). The same logic can be extended to the internal nodes of a bifurcating phylogeny, allowing a total of $n - 1$ independent contrasts to be formed from n given data points (Figure 5.3). Felsenstein's seminal contribution was to provide a method for computing these contrasts (Felsenstein, 1985).

The general principle of Felsenstein's method is easy to explain, but the details of the recursive algorithm by which it operates are not. There are two key complications. First, it is necessary to estimate values of the characters for the internal nodes of the tree, so that the differences between them can be calculated. Second, it is necessary to estimate the variance for each contrast, so that they can be rescaled to have the

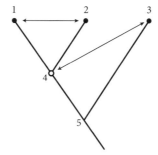

Figure 5.3 Given a bifurcating phylogeny with n tips, it is possible to form a total of $n - 1$ independent contrasts using the values of a character measured at the tips of the tree. For the simple case shown here, the difference in the values at nodes 1 and 2 is obviously independent of the difference in the values at nodes 3 and 4. Because values can only be measured directly at tip nodes (closed circles), it is necessary to estimate a value for the internal node (open circle) in order to form the second contrast. Hence, although the second contrast in principle compares the values of the character at nodes 3 and 4, this is achieved in practice by comparing the measured value of the character at node 3 with a (weighted) average of the measured values of the character at nodes 1 and 2. The root node (5) does not feature in the computation of these contrasts. Further standardization of the contrasts is required if they are to have the same variance.

same variance. Felsenstein's algorithm computes the necessary estimates by assuming a Brownian model of trait evolution. The resulting contrasts are expected to be normally distributed with unit variance and expected mean zero, provided that the assumed phylogeny and assumed model of trait evolution are correct. The contrasts that are computed by Felsenstein's method are therefore expected to satisfy the assumptions of ordinary least squares, even though the original data do not, provided that the relationship between the original variables is indeed linear.

Because the expected value of the contrasts is zero, the association between the characters must be estimated using an uncentred correlation coefficient rather than the usual Pearson's correlation coefficient (e.g. Rohlf, 2001). The estimated correlation coefficient can then be used to estimate the slope of the relationship (Felsenstein, 1985). Felsenstein's original method gives no estimate of the intercept of the relationship, because information about the actual values of the characters is discarded in the process of forming the contrasts. However, the intercept can be computed by estimating the numerical values of the characters at the root node, and then solving for the line that passes through this point (Garland, 2000).

5.4.2 Generalized least squares

Generalized least squares is a standard statistical technique that deals with violations of the ordinary least squares assumption of independent and identically distributed error (see e.g. Rencher and Schaalje, 2008). Like Felsenstein's method, it can be expressed as a linear transformation of the original data that makes the transformed data conform to the assumptions of ordinary least squares. The key difference is that whereas the transformation in Felsenstein's method is described by a recursive algorithm, the transformation in generalized least squares is described by a matrix operation. This transformation is uniquely defined if the error covariance matrix can be specified to within an unknown constant representing the error variance (Box 5.1). The only point at which evolutionary information enters the frame in the generalized least squares approach is in the specification of this error covariance structure.

One way of specifying the error covariance structure is to assume that the measured characters evolve according to some explicit model of trait evolution. If a Brownian model is assumed, and if the traits are linearly related, then results obtained using generalized least squares will be identical to results obtained using Felsenstein's method of independent contrasts. In fact, for this special case, the two methods are equivalent (Rohlf, 2001). Other models of trait evolution might also be assumed, but as we have discussed already, in a generalized least squares model it is only necessary to make detailed distributional assumptions about the error. The assumption that the measured traits evolve according to a Gaussian model of evolution is sufficient to allow the use of generalized least squares, when coupled with the assumption that the traits are linearly related, but it is certainly not a necessary condition for this. We return to this point below, but will assume for the time being that the error covariance matrix is known to within an unknown constant.

Box 5.1 The method of generalized least squares.

Our aim is to fit a linear model of the form

$$Y = X\beta + \epsilon$$

where the vector Y contains the n observations of the response variable, and where X is a full-rank $n \times (k + 1)$ matrix containing the k explanatory variables and a column of ones. The vector β contains the model parameters. The vector ϵ is the error, with expected mean $E(\epsilon) = 0$ and covariance matrix $\mathrm{cov}(\epsilon) = \sigma^2 V$, where σ^2 is the unknown error variance. In the special case that $V = I$, where I is the identity matrix, the Gauss-Markov theorem states that the ordinary least squares estimator $\hat{\beta} = (X'X)^{-1} X'Y$ is the best linear unbiased estimator of β (see e.g. Rencher and Schaalje, 2008). The covariance matrix for $\hat{\beta}$ is then given by $\mathrm{cov}(\hat{\beta}) = \sigma^2 (X'X)^{-1}$. In a typical cross-species comparative analysis, $V \neq I$ because of phylogenetic non-independence. In this case, the ordinary least squares estimator $\hat{\beta}$ remains unbiased, but is no longer best in the sense of having minimum variance among all linear estimators. Furthermore, its true variance will be underestimated by $\sigma^2 (X'X)^{-1}$, which will bias any significance tests. This is the motivation for using the generalized least squares estimator, which we now derive.

We will assume that V is a known positive definite matrix, which implies that there exists a non-singular lower triangular matrix P such that $V = PP'$. This factorization is known as the Cholesky decomposition (see e.g. Horn and Johnson, 2013), and is included as a function in most mathematical software. Multiplying both sides of the original model by P^{-1}, which is another lower triangular matrix, we obtain the transformed model

$$Z = U\beta + \delta$$

where $Z = P^{-1}Y$, $U = P^{-1}X$, and $\delta = P^{-1}\epsilon$. Note that the model parameters contained in β are unaffected by this transformation, although our estimates of them will be. The expected value of the error in the transformed model is $E(\delta) = E(P^{-1}\epsilon) = P^{-1}E(\epsilon) = P^{-1}0 = 0$. Because $E(\epsilon) = E(\delta) = 0$, it follows by definition that $\mathrm{cov}(\epsilon) = E(\epsilon\epsilon')$ and $\mathrm{cov}(\delta) = E(\delta\delta')$. We may therefore rewrite $\mathrm{cov}(\delta)$ as

$$\mathrm{cov}(\delta) = E(\delta\delta') = E\left(P^{-1}\epsilon \left(P^{-1}\epsilon\right)'\right)$$

$$= E\left(P^{-1}\epsilon\epsilon' (P')^{-1}\right) = P^{-1}E(\epsilon\epsilon') (P')^{-1}$$

$$= P^{-1}\sigma^2 V (P')^{-1} = \sigma^2 P^{-1}V (P')^{-1}$$

$$= \sigma^2 P^{-1}PP' (P')^{-1} = \sigma^2 I$$

which shows that the errors in the transformed model are independent and identically distributed because $\mathrm{cov}(\delta) = \sigma^2 I$. The Gauss-Markov theorem therefore implies that the ordinary least squares estimator $\hat{\beta}$ will be the best linear unbiased estimator of β if the transformed variables Z and U are substituted for the untransformed variables Y and X. This yields the generalized least squares estimator $\hat{\beta}^* = (U'U)^{-1} U'Z$, the covariance matrix of which is given by $\mathrm{cov}(\hat{\beta}^*) = \sigma^2 (U'U)^{-1}$. Significance tests made using this generalized least squares estimator will be unbiased, provided that the matrix V is specified correctly at the outset.

Given this knowledge, the original data can be transformed by factoring the matrix describing the error covariance structure into a pair of lower and upper triangular matrices, and multiplying the original data by the inverse of the first of these matrices (Box 5.1; see also Butler et al., 2000). The result of this transformation is that the error in a linear model fitted to the transformed data will be independent and identically distributed—assuming, of course, that the error covariance structure was correctly specified at the outset. Data that have been transformed in this way can therefore be analysed using any of the usual ordinary least squares techniques. This, in a nutshell, is the method of generalized least squares.

The transformation that is used in generalized least squares can be applied to categorical, as well as continuous, explanatory variables, provided that any categorical variables are coded as dummy variables (i.e. as $k - 1$ columns containing ones and zeros, where k is the number of levels of the categorical variable). This is an important advantage over Felsenstein's method, which deals only with continuous characters that are assumed to evolve according to a Brownian model of trait evolution. Clearly, a discrete character cannot be thought of as evolving according to a Brownian model—or indeed any other Gaussian model—of trait evolution. This brings us back to the important question of how we are to specify the error covariance structure.

5.4.3 Specifying the error covariance structure

The error in a fitted relationship will usually be due to a host of important but omitted variables. This means that its true covariance structure is not only unknown, but unknowable. Fortunately, it is usually possible to estimate a few parameters of the error covariance structure from the data. This contrasts with the approach sometimes used in comparative biology of pretending that the true error covariance structure is known exactly on the basis of some assumed model of trait evolution. It is most unlikely that the error covariance structure will be correctly specified by assuming a rigid evolutionary model, and in the worst case, it is actually possible to introduce error covariance in this way. Fortunately, the consequences of incorrectly specifying the error covariance structure are not too severe from the perspective of parameter estimation, because the least squares estimator remains unbiased (Rohlf, 2006). However, the estimator will not have minimum variance if the error covariance structure is incorrectly specified, and tests of the significance of the model parameters will therefore be biased.

A much better approach is to estimate one or two parameters of the error covariance structure at the same time as fitting the parameters of the relationship (e.g. Grafen, 1989; Díaz-Uriarte and Garland, 1996; Martins and Hansen, 1997; Pagel, 1997; Díaz-Uriarte and Garland, 1998). If a structure is assumed that allows as a special case for the possibility that there is no error covariance, then there should be little risk of introducing error covariance where none existed previously. Furthermore, the resulting estimators will remain unbiased, and are guaranteed to have lower variance than they would otherwise have had if a rigid error covariance structure had been assumed from the same family of error covariance structures. Unfortunately, significance tests become approximate when one or more parameters of the error covariance structure are estimated from the data, because the calculated F-statistic is no longer expected to follow an exact F-distribution (see e.g. McCulloch et al., 2008). We regard this as a small price to pay for the benefits that are gained.

Even if we estimate one or two parameters of the error covariance structure from the data, we will still need to assume a family of possible structures. Unfortunately, all that we can say from first principles is that the error covariance in a comparative analysis must be structured hierarchically along phylogenetic lines. This can be seen by treating the branches of the phylogenetic tree as representing all of the possible sources of error variance that are accumulated through evolution. For convenience, the length of each branch is taken to represent the total amount of error variance that accumulates along it. Where there are unresolved polytomies, these can be dealt with by assuming an arbitrary branching structure with zero branch lengths for the affected species (Rohlf, 2001). Measurement error is not expected to be shared by related species, so only contributes to branch length at the tips; equation error is expected to accumulate at all levels of the phylogeny. The error variance for each data point is therefore the total path length traced from root to tip through the tree, while the error covariance for each pair of data points is just the total path length that is shared in common between them. The hierarchical structure of the error covariance matrix then follows naturally from the hierarchical structure of the phylogeny.

Further detail of the error covariance structure cannot be specified from first principles without assuming an explicit evolutionary model. For example, if we assume a simple Brownian model of trait evolution, then the variance in a character becomes proportional to the length of time over which it has been evolving. This is too narrow a view, however, because what we are concerned with in general is the covariance of the equation error, and not the covariance of the characters that are contained in the model. The covariance of the equation error unquestionably depends upon the processes by which traits evolve, but it also depends upon the form of the relationship between traits, which will often be unknown. It follows that any method of assigning branch lengths is bound to be more or less arbitrary with respect to reality (Grafen, 1989). This is both a blessing and a curse. On the one hand, it means that we should not worry unduly if we lack information on divergence times, say, because it is far from obvious how the variance of the equation error should relate to evolutionary time. On the other hand, it leaves us with an arbitrary choice to make between the panoply of other methods that have been proposed for assigning branch lengths. The most important thing, therefore, is that we allow ourselves flexibility to distort the branch lengths when fitting the model.

In the absence of detailed information on the divergence times of all of the species in our dataset, we have chosen to use the method of branch length assignment that was advocated by Grafen (1989). Grafen suggests associating each node with a 'height' that is one less than the number of species with which it is associated, divided by one less than the total number of species in the tree. Species nodes are therefore assigned a height of zero, while the root of the tree is assigned a height of one. Grafen recommends raising the heights that are assigned to each node to the power of a parameter ρ that is fitted at the same time as the parameters of the relationship, using maximum likelihood techniques. This transformation has the effect of squashing or stretching the branch lengths to varying degrees at the roots or tips, and in the special case that $\rho = 0$ it allows for the possibility that there is no error covariance in the original data.

Other methods of assigning and distorting branch lengths are available (e.g. Martins and Hansen, 1997; Pagel, 1997). However, if the results of a comparative analysis are strongly influenced by the method that is used to assign and distort branch lengths, then those results ought obviously to be treated with some caution. In the analyses that we present in this book, we therefore flag any statistically significant results that would not have been significant, or which would have differed in their sign, had no attempt been made to control for phylogeny.

5.5 The comparative method in practice

In the final section of this chapter, we use the scaling of wing area with body mass in birds as a practical illustration of the use of generalized least squares. Wing area is expected to increase as the two-thirds power of body mass unless shape or density change systematically with body size, which means that wing loading is expected to increase as the one-third power of body mass under the null hypothesis of isometry. This is not a particularly interesting hypothesis in itself, but wing loading is one of the key parameters affecting flight performance, and it is therefore important to control for the scaling of wing area with body mass when testing for the effects of selection upon wing loading as we do in Chapter 7. Measurements of wing area are much more prone to measurement error than are measurements of body mass. It is therefore most reasonable to regress log wing area on log body mass, bearing in mind that we are thereby assigning all of the equation error, as well as all of the measurement error, to wing area (see Chapter 4).

We have already shown with reference to auks (Alcidae) that there is a high degree of phylogenetic non-independence in the relationship between wing area and body mass in birds. We will therefore need to control for phylogeny by using the method of generalized least squares to fit the regression. The effects of this phylogenetic error covariance are illustrated for all of the taxa in Figure 5.4, which displays the residuals of the ordinary least squares regression of log wing area on log body mass alongside the phylogeny of the species to which they refer. The residuals are strongly patterned along phylogenetic lines. For example, besides the cluster of negative residuals for auks that we have already discussed, there is another obvious cluster of negative residuals for fowl (Galloanserae), and an equally obvious cluster of positive residuals for broad-winged raptors (Accipitriformes). This phylogenetic error covariance will obviously need to be taken account of in any relationship that we fit to the data, which means that we will need to specify its expected structure using the methods described in the previous section.

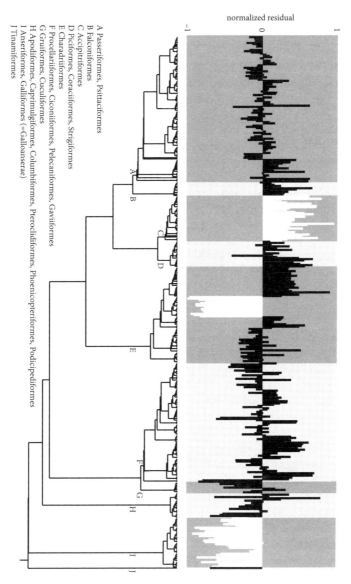

Figure 5.4 Bar chart showing normalized residuals from the ordinary least squares regression of log wing area on log body mass. The residuals have been normalized so that they fit on a scale from –1 to 1, and are shown alongside the phylogeny, with background shading delimiting clades A-J. Branch lengths have been distorted using the methods in Grafen (1989) at the same time as the parameters of the relationship are fitted. The branch lengths are therefore proportional to the amount of error variance that is estimated to have accumulated along them in light of the data. Although the tree may look to be poorly resolved in places, this is an artefact of the scale at which the branch lengths have been displayed. The groups of residuals that are highlighted in white correspond to the residuals for Accipitriformes, Alcidae, and Galloanserae. The clustering of residuals of the same sign within each of these clades indicates strong phylogenetic error covariance.

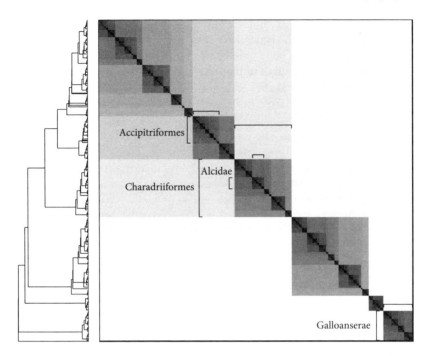

Figure 5.5 Graph showing the strength of the expected variance and covariance of each element or pair of elements, for the estimated error covariance matrix in the generalized least squares regression of log wing area on log body mass. The elements of the matrix are ordered so that members of the same clade are represented by adjacent rows and columns (see phylogeny below). The darkness of the grayscale shows the strength of the expected covariance for each element of the matrix. See text for further discussion.

The hierarchical nature of the error covariance structure is illustrated graphically in Figure 5.5. Error covariance matrices are always symmetric matrices, with the error variances of the data points running down the leading diagonal. The error covariances for the corresponding pairs of data points are contained in the off-diagonal elements. For example, an off-diagonal element in the ith row and jth column of the matrix gives the expected error covariance for the ith and jth data points. We have represented the error covariance matrix as an image in Figure 5.5, where the darkness of the pixels shows the strength of the expected variance or covariance estimated using Grafen's method of branch length assignment and distortion (Grafen, 1989). The rows and columns of the matrix are ordered so that members of the same clade are represented by adjacent rows and columns. Each square block corresponds to an identifiable clade in the phylogeny, and the hierarchical structure of the error covariance is clearly visible in the nesting of the square blocks running along the leading diagonal. For example, the block for Alcidae is nested within the block for Charadriiformes.

If we are successful in controlling for phylogeny, then no error covariance should remain in the residuals of the generalized least squares analysis. Unfortunately, because the transformed data points are computed by pooling information from the original data points, the residuals of the generalized least squares analysis cannot be associated with individual branches of the phylogeny, as was done for the residuals of the ordinary least squares analysis in Figure 5.4. Instead, we have plotted the sample covariance matrix of the residuals, where what we are looking for ideally is a matrix that shows no patterning. The image of the matrix of residual products for the untransformed data has a strong tartan-like pattern (Figure 5.6A), but the matrix of residual products for the transformed data shows no strong pattern (Figure 5.6B). This is a good indication that the generalized least squares transformation has been effective in removing the error covariance.

Controlling for phylogeny has rather little effect upon the estimate of the slope of log wing area on log body mass, which is 0.673 for the ordinary least squares regression and 0.668 for the generalized least squares regression. It is worth recalling that these two estimates of the slope are both unbiased estimates of the same underlying parameter (Rohlf, 2006). However, we expect the generalized least squares estimate to have lower variance than the ordinary least squares estimate, and hence to be more reliable. As it happens, the slope estimate of 0.668 from the generalized least squares analysis is almost exactly equal to the slope of 0.667 that would be expected under

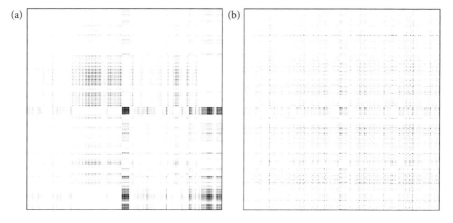

(a)   (b)

Figure 5.6 Sample covariance matrices for the residuals from the ordinary least squares (a) and generalized least squares (b) regressions of log wing area on log body mass. The elements of the matrix in (a) are ordered so that members of the same clade are represented by adjacent rows and columns, as in Figure 5.5. The elements of the matrix in (b) are ordered similarly following transformation of the data, but are no longer associated with single species. The darkness of the grayscale shows the relative magnitude of the sample covariance for the positive elements of the matrices. We have not displayed the negative elements of the matrices, because there is no reason to expect negative error covariance along phylogenetic lines in this example. See text for further discussion.

the null hypothesis of isometry. It is certainly statistically indistinguishable from this (95% confidence interval: 0.641, 0.695). The predictive power of the model is appropriately high ($R^2 = 0.84$), so we conclude that there is no evidence that selection upon wing loading varies systematically with body size.

5.6 Conclusions

When estimating a relationship between variables, phylogeny manifests itself through the effects of shared variation in important but omitted variables. Phylogenetic non-independence is nothing more, or less, than the result of fitting a model with equation error. Importantly, the distribution of this equation error will depend as much upon the form of the underlying relationship between the variables as it will upon the detailed processes by which the variables evolve. Furthermore, the equation error will obviously change if we add more, relevant terms to the model. The extent of the phylogenetic error covariance is therefore contingent upon the relationship that we fit, and it follows that we should not necessarily expect it to be straightforwardly related to any particular model of trait evolution. Fortunately, the method of generalized least squares is well able to deal with this eventuality, and we recommend its use in any comparative analysis involving a continuous response variable.

The emphasis that is usually placed upon relating the apparent influence of phylogeny to an explicit model of trait evolution is understandable insofar as those processes of trait evolution are of intrinsic interest to evolutionary biology. However, this emphasis is unnecessarily restrictive if the primary aim of using a phylogenetically controlled comparative method is to eliminate a statistical problem with the data. In this case, we are probably better off estimating some of the parameters of the error covariance structure from the data at the same time as fitting the relationship. In order to do this, we will need to make use of the fact that the error covariance structure will be structured along phylogenetic lines, so as to specify a possible family of error covariance structures. This is already an evolutionary model of sorts, and we should not be too quick to reach for further, more detailed, evolutionary assumptions unless it is absolutely necessary to do so.

In Chapter 7, we use the comparative method to test whether there is any evidence that the flight morphology of birds has adapted in response to specific ecological and behavioural selection pressures. Before doing so, however, we use dimensional analysis and simple physical reasoning to develop robust and testable predictions about how flight morphology is expected to respond to selection for different aspects of flight performance. This is the focus of Chapter 6.

6

Form and function in flight

6.1 Introduction

In the preceding chapters, we developed a suite of tools for exploring evolutionary questions in biomechanics. In this and the following chapters, we use these tools to make and test predictions about how natural selection is expected to shape the flight morphology of different species of bird. In so doing, we begin to flesh out some of the general conclusions about optimization that we drew in Chapter 2 in the context of our 'drowning landscape' model of adaptive evolution. There are at least four reasons why bird flight is a good example to consider to this end. First, the obvious similarities in wing design that exist among birds, bats, and pterosaurs are classic examples of convergent evolution. Second, theoretical understanding of aerodynamics is backed up by a wealth of empirical experience in aeronautical engineering. Third, the morphology, behaviour, and ecology of birds are exceptionally well known and documented. Fourth, there is a rich history of using the principles of flight mechanics to make testable predictions about bird flight (see especially Pennycuick, 1978; Hedenström, 2003).

In this chapter, we use the tools of dimensional analysis that we presented in Chapter 3 to derive robust, general conclusions about how flight morphology affects flight performance. We analyse the effects of flight morphology upon several different aspects of flight performance, including peak aerodynamic force production, steady glide speed, steady sink rate, aerodynamic power requirements, and aerodynamic efficiency. Clearly, the ecological relevance of these difference aspects of flight performance will vary according to the ecology and behaviour of the species in question. For example, having a low sink rate is expected to be of particular ecological relevance to soaring birds. To relate the rather loose notion of ecological relevance back to the more rigorous definitions of selective advantage and relative fitness in Chapter 2, this amounts to asserting that sink rate will be weighted more strongly in the fitness function of a soaring bird than in the fitness function of a non-soaring bird, all else being equal. Hence, we may reasonably refer to soaring species as being under selection 'for' low sink rate, in the sense that we expect soaring species to have a lower sink rate than non-soaring species, all else being equal.

In this chapter, we use the same reasoning to derive a set of simple directional predictions about how the flight morphology of birds is expected to look under selection for specific aspects of flight performance that we identify as being of particular

Evolutionary Biomechanics. Graham Taylor & Adrian Thomas.
© Graham Taylor & Adrian Thomas 2014. Published 2014 by Oxford University Press.

relevance to species having particular ecological or behavioural characteristics. The resulting predictions are not new (see e.g. Pennycuick, 1975, 2008; Norberg, 1990; Rayner, 1988), but they have rarely been derived from first principles. Furthermore, the results that have been used to make such predictions previously often rest implicitly or explicitly upon detailed aerodynamic theories, such as lifting-line theory or actuator disc theory, that require some detailed and often unrealistic physical assumptions (e.g. Pennycuick, 2008). The simpler approach that we use here permits us to make qualitatively identical predictions from first principles, with much less restrictive assumptions. In this way, we are able to make our directional predictions as general as they possibly can be, while also avoiding the need to explain detailed aerodynamic theory. For those readers who prefer a more explicit analytical treatment, we state the results of classical wing theory briefly at the end of this chapter to demonstrate that the use of more detailed aerodynamic theory leads to the same directional predictions (Box 6.1). We test these directional predictions in Chapter 7, making use of the comparative dataset for birds and the method of generalized least squares that we introduced in Chapter 5.

6.2 Scale effects in bird flight

A successful dimensional analysis requires some intuition of the variables that are likely to be important in the phenomenon at hand, and of those that can be safely ignored. Unfortunately, the behaviour of fluids is often quite counterintuitive. For example, even that most intuitive of intellects, Leonardo da Vinci, supposed that a beating wing generated lift by pressing on the air faster than it could escape, thereby compressing the air beneath (Anderson, 1997). We now know that small changes in pressure transmit through a fluid at the speed of sound (e.g. Tritton, 1988), so air is effectively incompressible for objects moving much slower than this. The highest reliable speed measurements for any flying animal are those recorded from a gyrfalcon (*Falco rusticolus*), clocked at $58 \, \text{ms}^{-1}$ during a hunting stoop (Tucker et al., 1998). This is about 0.17 times the speed of sound, and variations in air density due to compressibility are less than 1% in this range (Lighthill, 1986). It is clear, therefore, that we can safely neglect the effects of compressibility in the flight of birds.

Much less obviously, it is often possible to neglect the effects of viscosity, although we will need a quantitative criterion for deciding when it is reasonable to do so. The relevant quantity is the Reynolds number (Re), which emerges straightforwardly from a dimensional analysis of the problem. Suppose that we are interested in predicting the aerodynamic force (F) on a wing with some characteristic length scale (l), moving through a fluid at some constant speed (U). There are only two ways in which a fluid can transmit force to a solid: through pressure, which depends upon the fluid's density (ρ), and through friction, which depends upon the fluid's viscosity (μ). If we ignore shape effects for the time being, then these are all of the variables that we will require to undertake a dimensional analysis of scale effects in animal flight. As there are $n = 5$ dimensional variables, of which $k = 3$ have independent dimensions

Table 6.1 Variables to be considered in a dimensional analysis of the forces on an animal moving steadily through a fluid

Variable	Symbol	Dimension
aerodynamic force	F	MLT^{-2}
speed	U	LT^{-1}
fluid density	ρ	ML^{-3}
fluid viscosity	μ	$\mathrm{ML}^{-1}\mathrm{T}^{-1}$
length	l	L

(Table 6.1), the Pi theorem states that we can express the relationship between them in terms of $n - k = 2$ dimensionless products (see Chapter 3).

We may write the first dimensionless product as $\rho^{w}U^{x}l^{y}F^{z}$ where w, x, y, and z are numerical exponents for which we must solve. Setting $z = 1$ arbitrarily, it is easy to show that this leads to the dimensionless force $F/(\rho U^{2}l^{2})$. Taking ρ, l and U as repeating variables which we include in both dimensionless products, we may write the second dimensionless product as $\rho^{w}U^{x}l^{y}\mu^{z}$, where we are again left to solve for w, x, y, and z. Arbitrarily setting $z = -1$, we are led to the product $\rho l U/\mu$, which is the familiar Reynolds number Re. Typically, the characteristic length scale l is taken to be the wing chord (i.e. the mean distance between the leading and trailing edges of the wing), in which case we have the chord Reynolds number. We have therefore shown that scale effects in the steady flight of birds are governed by a dimensionless equation connecting the dimensionless force $F/(\rho U^{2}l^{2})$ to the Reynolds number Re $= \rho l U/\mu$. We may summarize this by writing

$$\frac{F}{\rho U^{2}l^{2}} = \Phi\,(\mathrm{Re}) \tag{6.1}$$

where Φ is some undetermined, and possibly quite complicated, function. Experience shows that it may be reasonable to treat a fluid as inviscid when Re $> 10^{2}$. This is not because viscous effects are unimportant, but rather because they are effectively confined to a thin boundary layer enveloping the animal, in what is termed an attached flow (see Figure 6.1 for explanation).

An important exception arises if the boundary layer becomes separated from the wing's surface, as it does if the wing is inclined at too high an angle to the flow. Boundary layer separation is likely to be avoided under steady conditions, however, because the stall that occurs on a conventional wing results in a substantial loss of useful aerodynamic force under steady conditions. As a good first approximation, we may therefore drop fluid viscosity μ from the list of important variables in Table 6.1, and may consequently drop the Reynolds number from Eq. 6.1. This amounts to saying that the dimensionless force $F/(\rho U^{2}l^{2})$ does not vary greatly with Reynolds number in steady flows for which Re $> 10^{2}$. This is an exceedingly useful simplification, because birds operate at Re $> 10^{2}$, which means that any conclusions we now draw

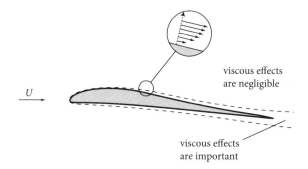

U

viscous effects
are negligible

viscous effects
are important

Figure 6.1 Diagram of the classical boundary layer approximation. Fluid adjoining a solid surface, such as a wing, is dragged along with it as a result of friction. The fluid's viscous tendency to resist shear results in a relative velocity gradient normal to the surface. The inner region of fluid in which the shear stresses are significant is known as the boundary layer (its thickness, shown by the dashed lines, has been greatly exaggerated in this figure). Because the boundary layer is thin relative to the airfoil, it only contains significant pressure gradients parallel to the surface, and these do not apply force to the airfoil. Pressure normal to the surface is impressed upon the boundary layer, so if the pressure forces are much larger than the friction forces, which is true at high Reynolds numbers (Re), then it can be a reasonable approximation to ignore the boundary layer flow altogether, and to treat the fluid as inviscid. Experience shows that this approximation is valid when $Re > 10^2$, as it is for even the smallest birds.

about the gross effects of flight morphology can be assumed to be scale-invariant. We will account for some of the more subtle effects of Reynolds number later, when modelling how friction drag affects fine soaring performance in Chapter 8.

6.3 Form and function in bird flight

We have ignored shape effects so far, but must obviously bring them into the picture if we are to establish how flight morphology affects flight performance. The single most relevant length scale for flight is the wingspan (b), because this is the morphological variable which determines the mass flow of air that the wing encounters as it travels forward. Wing area (S) is also important aerodynamically, because it determines the area over which fluid dynamic pressure acts to apply force. We may immediately combine these two variables to form a new dimensionless variable, b^2/S. This is an important dimensionless shape parameter known as the aspect ratio ($\mathcal{R}$), and is a measure of the narrowness of the wing (Figure 6.2).

The aerodynamic forces on a wing also depend upon its inclination to the airflow. This inclination is called the angle of attack (α), and is already dimensionless. To a first approximation, then, the dimensionless force on a bird is expected to be a function of aspect ratio and angle of attack. We may summarize this by writing

$$\frac{F}{\rho U^2 S} \approx \Phi\left(\mathcal{R}, \alpha\right) \tag{6.2}$$

where Φ is some undetermined function. We could, of course, add an arbitrary number of other dimensionless shape parameters as arguments of Φ, depending upon how much detail we require. These might include, for example, parameters describing the shape of the airfoil, or the relative frontal area of the body. In flapping flight, we might also need to include certain dimensionless kinematic parameters, such as Strouhal number and reduced frequency (Taylor et al., 2003), to account for the unsteadiness of the airflow. However, the usual practice in aerodynamics is to replace the undetermined function Φ and all of its possible arguments by a single dimensionless parameter known as the aerodynamic force coefficient (C_F), which differs from Φ by a conventional factor of 2:

$$\frac{2F}{\rho U^2 S} \approx C_F. \tag{6.3}$$

Rearranging to get at the aerodynamic force then gives:

$$F = \frac{1}{2}\rho U^2 S C_F. \tag{6.4}$$

In the sections that follow, we manipulate this one basic equation to explore how flight morphology affects various components of flight performance. We focus our attention upon those morphological variables for which there are already extensive measurements in the literature. In practice, this means that we will only make predictions about form and function that involve body weight, wing area, or wingspan.

6.3.1 Selection for large transient forces

Eq. 6.4 shows that the aerodynamic force which an animal can produce scales linearly with wing area, so it is not surprising to find that wing area scales strongly with body mass in birds (see Figure 5.1). Other things being equal, however, we would also expect to find larger wing areas in species which need to generate large transient forces. We would expect this to be true, for example, of species that use aerial feeding and that need to produce large transient forces to turn, or of species that sometimes carry additional loads.

6.3.2 Selection for high glide speed

At equilibrium, the total aerodynamic force (F) must exactly balance the animal's weight (W), so we can substitute W for F in Eq. 6.4 and rearrange to get at the steady glide speed:

$$U = \left[\left(\frac{2}{\rho} \right) \left(\frac{W}{S} \right) \left(\frac{1}{C_F} \right) \right]^{1/2}. \tag{6.5}$$

The ratio of body weight to wing area (W/S) is called the wing loading. Evidently, the steady glide speed of an animal depends upon the square root of its wing loading

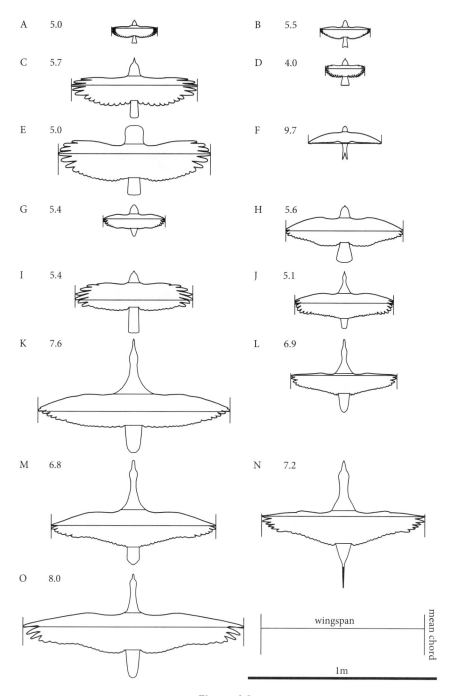

Figure 6.2

(W/S), so a bird can always glide faster by drawing its wings into its body to reduce their area. Although a high glide speed is obviously important for travelling long distances quickly, it is far from obvious how selection for glide speed should vary with flight ecology and behaviour. We return to this problem in Chapter 8.

6.3.3 Selection for low sink rate

Sink rate (U_s) depends upon glide angle (γ) and airspeed (U) as $U_s = U \sin \gamma$, which means that $U_s \approx U\gamma$ at the shallow glide angles associated with efficient flight. The aerodynamic force coefficient (C_F) is conventionally resolved into a lift coefficient (C_L) corresponding to the aerodynamic force component perpendicular to the relative airflow, and a drag coefficient (C_D) corresponding to the aerodynamic force component parallel to the relative airflow. It is easy to show graphically (Figure 6.3a) that the glide angle is related trigonometrically to the lift and drag coefficients as $\tan \gamma = C_D/C_L$, so that for small glide angles, $\gamma \approx C_D/C_L$. Finally, because $C_L = C_F \cos \gamma$, we can also make the small angle approximation $C_L \approx C_F$. Multiplying Eq. 6.5 by $\sin \gamma \approx \gamma$ to get at the sink rate, we therefore have

$$U_s \approx \left[\left(\frac{2}{\rho} \right) \left(\frac{W}{S} \right) \left(\frac{C_D{}^2}{C_L{}^3} \right) \right]^{1/2}, \tag{6.6}$$

which shows that the sink rate of a gliding bird depends upon the square root of its wing loading. Birds that are adapted for a low sink rate should therefore have a lower wing loading than birds that are not so adapted, other things being equal. Having a low sink rate is especially relevant to birds which soar in thermals and other kinds of updraft, because climbing is only possible in an updraft if the bird sinks slower relative to the air than the air rises relative to the ground (Figure 6.3b). Other things being equal, it follows that we would expect birds that soar in updrafts to have lower wing loading than birds that do not soar in updrafts.

Figure 6.2 Important morphological parameters for a range of different birds. Wingspan (b) is measured from wing tip to wing tip. Wing area (S) is measured over the area shown in grey, including the portion of body between the wings. Wing mean chord (c) is defined as wing area (S) over wingspan (b). The number to the left of each bird is its aspect ratio, which is a dimensionless shape parameter measuring the narrowness of the wings ($R = b^2/S = b/c$). A. Chaffinch *Fringilla coelebs*; B. European Greenfinch *Carduelis chloris*; C. Eurasian Jackdaw *Corvus monedula*; D. European Robin *Erithacus rubecula*; E. Tawny Owl *Strix aluco*; F. Common Swift *Apus apus*; G. Common Quail *Coturnix coturnix*; H. Stock Pigeon *Columba oenas*; I. Eurasian Sparrowhawk *Accipiter nisus*; J. Common Moorhen *Gallinula chloropus*; K. Common Eider *Somateria mollissima*; L. Hooded Merganser *Lophodytes cucullatus*; M. Mallard *Anas platyrhynchos*; N. Northern Pintail *Anas acuta*; O. Red-breasted Goose *Branta ruficollis*. Data from Thomas and Taylor (2001).

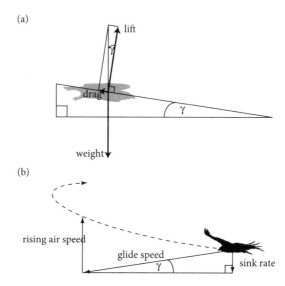

Figure 6.3 (a) At equilibrium, the cotangent of the glide angle (γ) is equal to the ratio of lift to drag. (b) The essence of thermal soaring is to sink slower, relative to the air in the thermal, than the air in the thermal rises relative to the ground: the dashed line shows the bird's trajectory. The climb rate is equal to the speed at which the air rises, minus the sink rate, and so is maximized by minimizing sink rate.

6.3.4 Selection for low power requirements

Gliding animals sink with respect to the surrounding air because some component of their body weight has to be directed along their flight path in order to balance drag. In steady level flight, a gliding animal's weight does work against drag at a rate equal to the rate at which potential energy is lost (i.e. weight times sink rate). In level powered flight, the same work against drag is done instead by the aerodynamic thrust. Hence, if the relative airflow over the wings in powered flight is similar to what it would be if the animal were gliding, which is the usual quasi-steady assumption, then the required aerodynamic power can be obtained by multiplying both sides of Eq. 6.6 by the animal's weight:

$$P \approx W \left[\left(\frac{2}{\rho} \right) \left(\frac{W}{S} \right) \left(\frac{C_D{}^2}{C_L{}^3} \right) \right]^{1/2}. \tag{6.7}$$

The total rate at which energy is consumed in flight is directly related to this aerodynamic power requirement.[1] Reduced aerodynamic power requirements will presumably be beneficial for any flying animal, but they may be especially important

---

[1] Eq. 6.7 has been deliberately constructed to eliminate airspeed (U), so as to clarify the effects of flight morphology. Nevertheless, variation in aerodynamic power requirements with airspeed is implicit in Eq. 6.7, because the drag coefficient (C_D) includes an induced drag component which increases with the lift coefficient. Other things being equal, a different operating lift coefficient is

Table 6.2 Variables to be considered in a dimensional analysis of the efficiency of momentum transfer in flight

Variable	Symbol	Dimension
momentum transfer rate	$\dot{p}$	MLT^{-2}
kinetic energy transfer rate	$\dot{E}$	ML^2T^{-3}
mass flow rate	$\dot{m}$	MT^{-1}
efficiency factor	e	dimensionless

to birds which fly long distances on a daily or annual basis by commuting or migrating. Eq. 6.7 shows that adaptations to reduce weight (W) or to increase wing area (S) will both decrease the energetic costs of flight. Other things being equal, we would therefore expect birds which migrate or commute to have lower wing loadings than birds that do not migrate or commute.

6.3.5 Selection for high aerodynamic efficiency

We have shown that an enhanced lift-to-drag ratio is expected to reduce sink rate and glide angle (Eq. 6.6), as well as reduce the aerodynamic power required for level flight (Eq. 6.7). We therefore expect that selection for high aerodynamic efficiency will act to increase the lift-to-drag ratio. A significant part of the drag produced by a real wing is induced drag (D_i), which is produced as an inevitable consequence of lift production. Newton's laws of motion tell us that the reaction to the lift force (L) is a transfer of momentum (p) to the wake at a rate $-\dot{p} = L$. This reaction transfers kinetic energy (E) to the wake, and it does so at a rate ($\dot{E}$) equivalent to the rate at which work is done by the induced drag (i.e. $\dot{E} = UD_i$).

The rates of momentum transfer and kinetic energy transfer have units of mass flow rate times velocity, and mass flow rate times velocity squared, respectively (see also Table 6.2). It is therefore straightforward to form a dimensionless product (e) by inspection, by dividing the squared rate of momentum transfer by the rate of kinetic energy transfer and the mass flow rate ($\dot{m}$) in the downwash:

$$e = \frac{\dot{p}^2}{2\dot{m}\dot{E}}. \tag{6.8}$$

The denominator of the expression for e includes a factor of 2, to cancel the factor of one-half that appears in the definition of kinetic energy. This is simply a convenience so that $e = 1$ in the special case that the downwash speed (v) is constant across the wing, because then $\dot{p} = \dot{m}v$ and $\dot{E} = \frac{1}{2}\dot{m}v^2$. It is easily verified that the dimensionless parameter e as defined here is equivalent to the span efficiency factor given by classical wing theory (see e.g. Katz and Plotkin, 2001).

---

required in order to fly at a different airspeed, and this means that the ratio C_D^2/C_L^3, and hence the total power requirement, also varies as a function of airspeed.

Box 6.1 Results of classical wing theory.

The following results assume that the wing is shaped so that the aerodynamic loading is distributed elliptically across the span, which results in an even downwash distribution and span efficiency factor $e = 1$. The other assumptions of the theory are quite abstract, but they are dealt with in most standard texts on aerodynamic theory (e.g. Katz and Plotkin, 2001). For wings of very low aspect ratio ($\mathcal{R} < 1$), slender wing theory predicts the lift coefficient as

$$C_L = \frac{\pi \alpha}{2/\mathcal{R}}$$

where α is the angle of attack. For wings of very high aspect ratio, lifting-line theory predicts the lift coefficient as

$$C_L = \frac{2\pi \alpha}{1 + 2/\mathcal{R}}$$

These asymptotic approximations have been widely used in the biomechanics literature (e.g. Thomas, 1993; Maybury et al., 2001), but neither gives accurate results at the intermediate aspect ratios of most flying animals. However, there is a higher-order asymptotic approximation available that is accurate to within 1% of the exact numerical solution for $\mathcal{R} \geq 2.55$ (Van Dyke, 1964). This predicts the lift coefficient as

$$C_L = \frac{2\pi \alpha}{1 + 2/\mathcal{R} + 16(\log(\pi \mathcal{R}) - 9/8)/(\pi \mathcal{R})^2}$$

where one of the coefficients in Van Dyke (1964) has been corrected after Kerney (1972). This higher-order approximation does not appear to be widely known, having been largely supplanted by numerical methods, but where a simple analytical prediction of lift coefficient is required, as is often the case in biomechanics, we would always recommend using it in favour of the usual approximation from classical lifting-line theory.

Lift cannot be generated without inducing drag, and the induced drag coefficient (C_{Di}) can be calculated from any of the predictions of lift coefficient above as

$$C_{Di} = \frac{C_L{}^2}{\pi \mathcal{R}}.$$

It is easy to confirm that this equation corresponds to the result that we obtained from momentum considerations for the ratio of lift to induced drag on a wing (Eq. 6.9) as follows. Making use of the identity $\mathcal{R} = b^2/S$ and rearranging yields

$$\frac{C_L}{C_{Di}} = \frac{\pi}{C_L} \frac{b^2}{S}$$

which differs from the right-hand side of the proportionality in Eq. 6.9 by a factor of $\pi/(4e)$ where $e = 1$ for an elliptically loaded wing. In fact, it is possible to achieve slightly lower induced drag coefficients using swept lunate planforms, but this non-linear effect makes relatively little difference to the magnitude of the forces produced.

The mass flow rate is equal to the density of the air (ρ) times the speed of the wing (U) times some characteristic area. Because the only significant length scale normal to the flow is the wing's span (b), it is reasonable to define this characteristic area as being proportional to span squared and to write the proportionality $\dot{m} \propto \rho U b^2$. Substituting this proportionality and the relations $\dot{p} = L$ and $\dot{E} = U D_i$ that we obtained earlier in Eq. 6.8 and making use of the identities $L = \frac{1}{2}\rho U^2 S C_L$ and $D_i = \frac{1}{2}\rho U^2 S C_{D_i}$, we can write

$$\frac{C_L}{C_{D_i}} \propto \left(\frac{4e}{C_L}\right)\left(\frac{b^2}{S}\right) \tag{6.9}$$

for the ratio of lift to induced drag. The results of classical wing theory stated in Box 6.1 show that the missing constant of proportionality in this relationship is equal to $\pi/4$ under the more detailed assumptions that this theory entails.

Equation 6.9 shows that the ratio of lift to induced drag varies linearly with aspect ratio ($\mathcal{R} = b^2/S$). The same result is also confirmed by the results of classical wing theory stated in Box 6.1. High-aspect ratio wings are therefore more efficient than wings of lower aspect ratio. Other things being equal, we would therefore expect selection for higher aerodynamic efficiency to result in higher aspect ratio wings. Aerodynamic efficiency is important for any flying bird, but it will be especially important in soaring birds, which need to avoid losing too much height as they cover ground, and therefore need to be efficient in order to achieve a shallow glide path. High aerodynamic efficiency is also expected to be especially important in migrant and commuting species, which will benefit strongly from the reduced induced power requirements that this confers (Eq. 6.7). We therefore expect birds that soar, migrate, or commute to have comparatively high aspect ratio wings.

6.4 Classical wing theory

Although we have derived all of the directional predictions above without having to make any of the detailed assumptions of classical wing theory, we will need to make use of the more detailed predictions of that theory later in Chapter 8. The key analytical results of classical wing theory are stated in Box 6.1 and are plotted in Figure 6.4 to show how the lift coefficient and induced drag coefficient vary with aspect ratio and angle of attack. These two graphs summarize the dynamical constraints upon aerodynamic force production in gliding animals: no real animal can depart far from these surfaces, which are a more-or-less inevitable outcome of the underlying fluid dynamics.

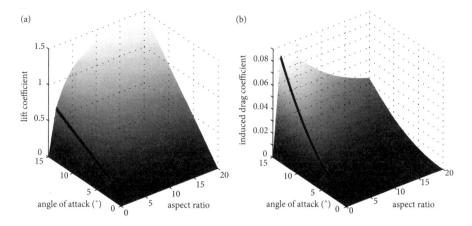

Figure 6.4 Surfaces plotting lift coefficient (C_L) and induced drag coefficient (C_{Di}) against aspect ratio ($\mathcal{R}$) and angle of attack (α) for wings with elliptic loading. The surfaces are predicted using classical slender wing theory in the range $0 \leq \mathcal{R} \leq 1.5$ and using a higher-order lifting-line theory (Van Dyke, 1964; Kerney, 1972) in the range $1.6 \leq \mathcal{R}$ (see Box 6.1). The model assumes attached flow, which can only be maintained at angles of attack up to approximately $\alpha = 15°$ on high aspect ratio wings; low aspect ratio wings may start to display flow separation at $\alpha \geq 5°$.

6.5 Conclusions

Our dimensional analysis of flight mechanics has allowed us to identify several specific morphological adaptations that we expect to find in birds under selection for specific components of flight performance. The effects of changes in flight morphology upon flight performance are summarized graphically in Figure 6.5, and allow us to make the following directional predictions about how we expect flight morphology to vary with flight ecology and behaviour:

(1) Selection for the ability to produce large transient forces is expected to manifest itself in adaptations to increase wing area. We therefore expect species that use aerial feeding, or that carry loads, to have comparatively large wing areas.
(2) Selection for low sink rate is expected to manifest itself in adaptations to reduce wing loading. We therefore expect species that soar in updrafts to have comparatively low wing loading.
(3) Selection for low power requirements in flapping flight is expected to manifest itself in adaptations to reduce wing loading. We therefore expect species that migrate or commute to have comparatively low wing loading.
(4) Selection for high aerodynamic efficiency is expected to manifest itself in adaptations to increase aspect ratio. We therefore expect species that soar, migrate, or commute to have comparatively high aspect ratio wings.

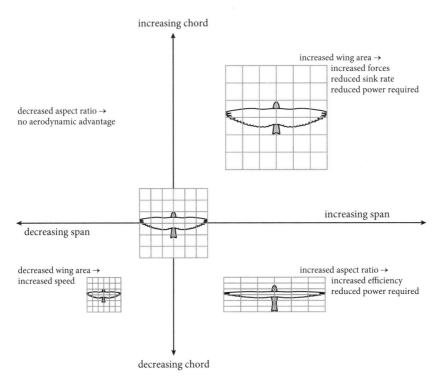

Figure 6.5 Theoretical effects of changes in flight morphology upon flight performance. The transformations described here are all simple stretches, but the grid behind each bird is intended to recall D'Arcy Thompson's classic descriptions of more general morphological transformations (Thompson, 1917).

Our aim in Chapter 7 will be to test these predictions empirically by using the phylo-genetically controlled comparative method to analyse the comparative dataset for avian flight morphology that we introduced in Chapter 5.

7

Adaptation in avian wing design

7.1 Introduction

In this chapter, we apply the theoretical and comparative tools developed in the previous chapters to the messy reality of biological data. We use the method of phylogenetically controlled generalized least squares that we introduced in Chapter 5 to test whether flight morphology varies with flight ecology and behaviour as expected under the directional predictions about form and function that we made in Chapter 6. The evolution of avian flight morphology provides a classic example for this sort of analysis, because the underlying theoretical predictions are derived from first principles from outside biology, and are backed up by over a century of intensive research in aeronautical engineering. So the biomechanical predictions are as firm as they can be, but the biological system is complicated enough to illustrate many of the hazards and opportunities for comparative analyses in evolutionary biomechanics.

We argued in the previous chapter that fitness will be a more strongly increasing function of certain components of flight performance in species which possess certain behavioural or ecological characteristics. For example, fitness is expected to increase more strongly with decreasing sink rate in birds which soar, than in birds which do not soar. Hence, because sink rate decreases with decreasing wing loading, we would expect wing loading to be lower in birds which soar, than in birds which do not soar, other things being equal. The qualification at the end of this sentence is important. For example, wing loading scales as the one-third power of body mass under the assumption of isometry, so the statement that wing loading is expected to be lower in soaring birds than in non-soaring birds only makes sense if we control for the effects of body mass. For this reason, we control for body mass when testing our directional predictions about variation in flight morphology. Trade-offs with other conflicting aspects of flight performance, such as aerodynamic efficiency, might also mask the effects of selection for low sink rate. We therefore test all of our directional predictions for any given morphological character simultaneously. We analyse the detailed consequences of trade-offs between conflicting aspects of flight performance in Chapter 8.

The theoretical predictions about wing design that we developed in Chapter 6 are only half of the story, because they need to be related to observable ecological or behavioural characters in order to be tested. To assess selection pressures on flight performance, we select ecological and behavioural characters that are linked

Evolutionary Biomechanics. Graham Taylor & Adrian Thomas.
© Graham Taylor & Adrian Thomas 2014. Published 2014 by Oxford University Press.

to plausible flight performance objectives, and that are so striking that they are used as identifying characters in general ornithological handbooks. Such characters are therefore available to be scored accurately and objectively by any researcher. This scoring is particularly straightforward for birds, because of the richness of the ornithological literature, but in principle similar sets of characters could be devised for any other biomechanical system. Defining an objective set of characters forces scrutiny of the observed biology, and can overturn some deep preconceptions, as is made clear in the descriptions of several of our ecological predictors below.

The convergence in the wing designs of many distantly related species is so striking that even quite fine details had been noticed decades before aeronautical engineering took off. We therefore begin this chapter by reviewing some previous analyses of the relationship between flight morphology, flight performance, and flight ecology in birds.

7.2 History

The wings of birds have been a classic example of convergent evolution since they were first discussed by Darwin. The adaptations of the wings of birds are discussed repeatedly in the *Origin*. For example, '*Can a more striking instance of adaptation be given than that of a woodpecker for climbing trees and for seizing insects in the chinks of the bark? Yet in North America there are woodpeckers which feed largely on fruit, and others with elongated wings which chase insects on the wing*' (Darwin, 1859). Darwin also noted many examples of convergent evolution of increased wing loading in species that swim to catch prey: '*Petrels are the most aerial and oceanic of birds, yet in the quiet Sounds of Tierra del Fuego, the* Puffinuria berardi [*Common Diving-Petrel* Pelecanoides urinatrix], *in its general habits, in its astonishing power of diving, its manner of swimming, and of flying when unwillingly it takes flight, would be mistaken by any one for an auk or grebe; nevertheless, it is essentially a petrel, but with many parts of its organisation profoundly modified.*' (Darwin, 1859).

Remarkably, Darwin's insight preceded the earliest recognition of the clear relationships between form and function in wing designs in the aeronautical literature. In a paper delivered to the first meeting of the Aeronautical Society of Great Britain in 1866, Wenham pointed out that '*it may be remarked that the swiftest-flying birds possess extremely long and narrow wings, and the slow, heavy flyers short and wide ones*' (Wenham, 1866). The same theme was picked up by aviation pioneer Otto Lilienthal, whose aim was explicitly biomimetic. As the title of his book *Bird Flight as the Basis of Aviation* (Lilienthal, 1889) makes clear, Lilienthal measured bird wing designs to establish the design parameters for his gliders. Lilienthal identified two general classes of efficient flight-design: a seabird model, and a land-bird model based on raptors and storks. The land-bird model had long broad wings, giving a moderate aspect ratio, low wing loading and slow flight. The seabird model had higher wing loading and long narrow wings, giving a higher aspect ratio and higher flight speed but a flatter glide angle (Lilienthal, 1889).

Substantial surveys of flight morphology were first collated by Müllenhoff (1885) for bats, by Demoll (1918) for insects, and by Magnan (1922), Poole (1938) and Hartman (1961) for birds. These datasets were augmented with new data, synthesized and analysed by Greenwalt (1962, 1975), using dimensional, aerodynamic and performance analyses. Perhaps most impressive in Greenwalt's work is his prescient recognition of the influence of phylogeny on the dimensional relationships amongst flying animals, which preceded the development of modern comparative methods by at least two decades. Greenwalt noted that the dimensional relationships between morphological characters varied in both slope and intercept between groups at species and at family levels. Greenwalt carefully considered these phylogenetic effects, concluding that: '*A plot of weight and wing area in logarithmic coordinates shows immediately that birds must be subdivided into broad classes or "models" if self-consistence of the data is to be achieved. By a process of cut and try, I have arrived at three "models", each differing substantially from the other in wing loading at a given total weight . . . It will be apparent that each "model" contains strange bed-fellows; the passeriform model, for example, includes the herons, the falcons, hawks, eagles and owls; the shorebird model includes the doves, parrots, geese, swans, two bustards and the single albatross for which we have data; the duck model includes grebes, loons and coots*' (Greenwalt, 1975). Greenwalt was able to describe variations in dimensional relationships between groups of birds, but his analyses were unable to address the selection pressures responsible for those patterns of variation.

There have been several efforts to relate adaptations in bird flight morphology to ecology and behaviour. For example, Rayner (1988), Norberg (1990) and Pennycuick (2008) have followed Greenwalt (1975) in using a least squares approach to examine the variation in dimensional relationships between members of different phylogenetic groups. More recent work has examined the adaptations of birds' wings in relation to their flight ecology, including the convergence between birds and bats (reviewed in Hedenström et al., 2009). This work has applied modern phylogenetically controlled techniques to specific isolated aspects of flight behaviour, such as flight speed on migration (Alerstam et al., 2007). Other work has analysed specific taxa—notably Pelecaniformes, in which the influence of flight style on wing morphology and on skeletal structure has been analysed in some detail (Brewer and Hertel, 2007; Simons, 2010). There has also been considerable interest, and indeed controversy, in relating skeletal structure to flight-styles with the aim of inferring the flight capabilities of fossil species (Wang et al., 2011; Chan et al., 2013). However, there has been no attempt to date to provide a rigorous quantitative analysis of how avian flight morphology adapts in response to ecological and behavioural selection pressures, which is the aim of this chapter.

7.3 Ecological predictors

The key advantage of biomechanical analyses is the ability to make a priori predictions of the kind set out in the previous chapter, based on the underlying physics, and independent of the detailed biology. In Chapter 6 we developed a set of directional predictions about the adaptations that we would expect to find in the flight morphologies of birds under selection for particular aspects of flight performance. These predictions are summarised in Figure 7.1, which is drawn using changes in the two orthogonal axes of wing morphology—span and chord—to drive changes in wing area and aspect ratio. The difficult problem is in linking the performance objectives that we have identified to observable behavioural or ecological characters. In rare cases, specific aspects of flight performance have a clear direct impact on fitness. For example, maximum climb rate is surely critical to a skylark (*Alauda arvensis*) pursued by a merlin (*Falco columbarius*); so much so that a lark confident of a superior climb rate sings to tell the merlin not to prolong the chase (Cresswell, 1994). Similarly, maximum climb angle determines the suitability of breeding sites for red-throated divers (*Gavia stellata*), which are unable to use many lakes because they simply cannot take off and climb fast enough to clear the surrounding trees and hills (Norberg and Norberg, 1971). However, such direct correlations between flight performance and survival or reproduction are rare.

Even with the most tightly defined mission, performance remains a compromise. For example, sailplanes are designed to achieve the maximum average cross country speed, and their designs are optimized through detailed, highly funded research. However that optimization is complex, because a sailplane not only has to glide efficiently at high speed, but also has to turn tightly while maintaining a low sink rate in thermal updrafts (see e.g. Thomas, 1999). Gliding at high speed requires high wing loading, and stiff, strong wings to resist flutter. Low sink rate requires high aspect ratio and low wing loading, which makes the wings prone to flutter. Hence, even in such highly specialized aircraft, the design represents a compromise between conflicting optimization criteria. The wing designs of birds must also be severely compromised—they have to feed and breed as well as fly—so the selection pressures on flight performance will be far more complicated than any single performance metric could reveal.

The multiple selection pressures that produce these trade-offs are so complex that they cannot be identified with any confidence from first principles. A taste of the simplest and most straightforward trade-offs is given in Figure 7.1. In the absence of any way to resolve these trade-offs theoretically, our approach is to analyse them by applying the comparative method to analyse the interaction between flight morphology and ecological characters that relate to flight performance. We score these ecological characters as binary variables, determined according to whether or not they are recorded in the ornithological handbooks as a character of each species. The mapping of these characters onto the aerodynamic predictions is hardly likely to be precise, but selecting characters conspicuous enough to be recorded in ornithological handbooks and using a binary scoring system of presence or absence for each character makes the characters as objective and reliable as possible.

For each species of bird, we scored each of the following eight characters using the multi-volume handbooks for the various regions of the world. Where possible,

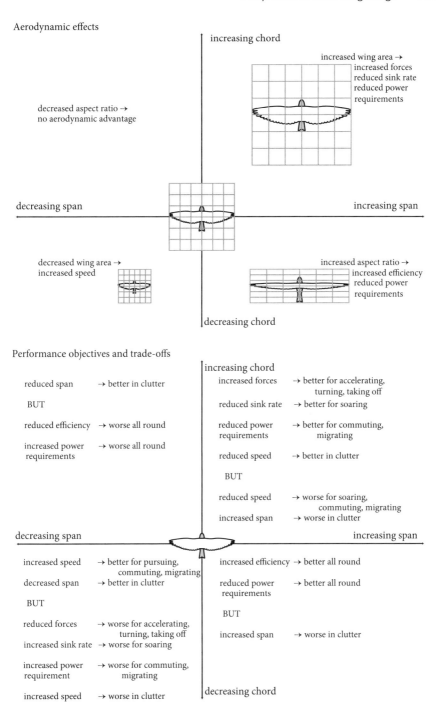

Figure 7.1 Predicted aerodynamic effects of wing morphology on flight performance and the associated performance objectives and trade-offs related to ecology and behaviour.

we made use of either the *Birds of the Western Palearctic*,[1] the *Handbook of Australian, New Zealand and Antarctic Birds*,[2] or the *Birds of North America*.[3] We used the *Handbook of the Birds of the World*[4] for any species not included in these regional handbooks. For detail on the hunting techniques of birds of prey, we referred to the *Raptors of the World* (Ferguson-Lees and Christie, 2001). In order to avoid making subjective judgements, we scored each character according to whether or not it was recorded at all, without regard to the relative importance of the character to a given species.

(1) *Exclusively pelagic use of soaring*
There are a number of birds which soar exclusively over the sea, and which presumably never make use of thermals. This category excludes those marine soaring species such as the snow petrel *Pagodroma nivea* and light-mantled albatross *Phoebetria palpebrata* which are recorded as soaring to great heights over land, presumably in thermals, and which therefore score '0' for exclusively pelagic use of soaring. All of the other Procellariiformes in our dataset score '1' for this character, except for the common diving-petrel *Pelecanoides urinatrix* and South Georgia diving-petrel *Pelecanoides georgicus*, which do not soar at all (Marchant and Higgins, 1990). A number of other species of soaring seabird that are not recorded as soaring over the land also score '1' for this category. In total, 77 species in our dataset (17%) make exclusively pelagic use of soaring. We predict that birds that soar over the sea should be under selection for high aerodynamic efficiency and low sink rate. Our aerodynamic predictions therefore suggest that they should have high aspect ratio and low wing loading.

(2) *Soaring over land*
Our definition of soaring over land is intended to include any soaring bird that makes use of thermals. We assumed that any species recorded as soaring over land would make use of thermals, and assumed that predominantly marine species did not make use of thermals unless specifically recorded as doing so. This category includes all of the Accipitriformes in our dataset, and a large number of other species that are recorded in the handbooks as soaring over land. Besides obvious examples such as the common crane *Grus grus*, there are a few surprises in this category, including the anhinga *Anhinga anhinga* which scores '1' for soaring over land. In total, 103 of our species (23%) soar over land. We predict that birds that soar over land should be under selection for high aerodynamic efficiency and low sink rate, but in contrast to birds that only soar over the sea, they should also be under selection for large flight forces to allow them to circle tightly in thermal

---

[1] Cramp (1977, 1980, 1983, 1985, 1988); Cramp and Brooks (1992); Cramp et al. (1993); Cramp and Perrins (1994a,b).

[2] Marchant and Higgins (1990, 1993); Higgins and Davies (1996); Higgins (1999); Higgins et al. (2006).

[3] Poole et al. (1992–2002).

[4] del Hoyo et al. (1992, 1996, 1997, 2005, 2006, 2009).

updrafts. Our aerodynamic predictions therefore suggest that they should have high aspect ratio, low wing loading, and large wing area.

(3) *Submerged aquatic feeding*

Our definition of submerged aquatic feeding encompasses all birds that are recorded as submerging completely to feed. Birds that do so habitually include species ranging from aerial plunge divers like the brown pelican *Pelecanus occidentalis* to surface divers like the great northern diver *Gavia immer*, which both score '1' for this category. Numerous other species submerge occasionally, such as the northern fulmar *Fulmarus glacialis*, which mostly seizes food from the surface, but occasionally uses pursuit-plunging to depths of over 4m (Cramp, 1977) so also scores '1' for use of submerged aquatic feeding. This is a nuanced category, because close relatives can differ markedly in their propensity to submerge. For example, the white pelican *Pelecanus onocrotalus* feeds while swimming on the surface and does not plunge dive (Cramp, 1977), so scores '0' for this category. In general, we scored species as using submerged aquatic feeding if they were recorded as using 'diving' or 'pursuit plunging' or any variant of those terms. In total, 120 of our species (27%) use submerged aquatic feeding. This should require smaller wings to reduce the buoyancy of the gas-filled feathers, and to cope with the higher density of water and the larger force-per-unit-area in those species which use their wings for propulsion underwater. Birds that use submerged aquatic feeding should therefore be under selection leading to adaptations for smaller wings and increased wing loading, but we can make no unequivocal prediction about the effect on aspect ratio.

(4) *Sally hunting*

Our definition of sally hunting includes sallying after aerial prey from the ground or a perch, as well as other forms of aerial fly-catching behaviours that are performed intermittently. For example, the spotted flycatcher *Muscicapa striata* habitually makes use of short sallies after aerial prey, so obviously scores '1' for this character. The tree pipit *Anthus trivialis* is recorded as occasionally taking insects after a short aerial pursuit from the ground (Cramp, 1988), so also scores '1' for use of manoeuvrable aerial feeding, even though this is presumably a less important selective pressure in this species. Its close relative the meadow pipit *Anthus pratensis* also takes insects, but is said never to fly after them (Cramp, 1988), and so scores '0' for this character. Key descriptions in the guide books include any behaviour described as 'sallying', 'still hunting', 'dash from a perch', 'fluttering pursuit', or 'short aerial pursuit'. In total, 106 of our species (24%) are recorded as sally hunting. The unifying theme is the need for high accelerations and large flight forces, and we therefore predict that sally hunters should have increased wing area relative to their body mass, and thus low wing loading.

(5) *Pursuit hunting*

Our definition of pursuit hunting includes predatory and piratical aerial pursuit behaviours, and other examples of continuous aerial hawking. For example, the swift *Apus apus* hawks continuously so obviously scores '1' for this category. The magnificent frigatebird *Fregata magnificens* obtains most of its food from the ocean surface, but engages in occasional kleptoparasitic behaviour of other birds in flight (Cramp, 1977), so less obviously scores '1' for pursuit hunting. The great majority of species score '0' for this character. We defined pursuit hunting as including any behaviour described as 'aerial hawking', 'extended chase', 'tail chase', 'stooping', 'piracy', 'kleptoparasitism', and 'aerial pursuit'. In total, 73 of the species in our dataset (16%) are recorded as using pursuit hunting. The unifying theme here is the need to generate large flight forces in order to turn quickly at high flight speeds. Large flight forces require large wing area relative to body mass or low wing loading, but high flight speeds require high wing loading. These are in direct conflict, so we can make no unequivocal directional prediction in this case.

(6) *Use of cluttered airspace*
This character is difficult to score objectively, so we only scored a species as 0 for cluttered airspace in the clearest cases. Birds which soar high over clutter (e.g. the griffon vulture *Gyps fulvus*), or birds which lead an essentially marine existence (e.g. the wandering albatross *Diomedea exulans*) obviously score '0' for use of cluttered airspace. Birds which exclusively inhabit steppes, open fields, desert, tundra, screes, moorland, or lagoon environments also score '0' for this character. We recorded a '1' for aerial clutter if there was any kind of clutter in the habitat at all. For example, uniquely among auks (Alcidae), the marbled murrelet *Brachyramphus marmoratus* nests on the limbs of old-growth trees and flies through gaps in the forest canopy in order to do so. It therefore scores '1' for use of cluttered airspace. In contrast, the magnificent frigatebird *Fregata magnificens* lands on the tops of trees or shrubs to nest, and therefore scores '0' for use of cluttered airspace. There are also quite a number of species that live in open environments at one time of the year, but which must contend with aerial clutter at other times, and these have all been scored as a '1'. A total of 214 (48%) of our species have to contend with aerial clutter. We predict that birds that fly in clutter will require large flight forces and the ability to fly slowly, and will therefore have large wing area and low wing loading.

(7) *Use of migration*
Our definition of migration excludes seasonal movements that are purely dispersive, altitudinal, or irruptive but includes any species for which all or part of the population migrates. For example, the arctic tern *Sterna paradisaea* is a trans-equatorial migrant, so obviously scores '1' for migration. The skylark *Alauda arvensis* makes no more than local movements in the south of its Palaearctic range but is wholly migratory in the north and east (Cramp, 1988) and so also scores '1' for migration. The nuthatch *Sitta europaea* is chiefly sedentary, making only dispersive or occasional irruptive movements and so scores '0' for migration. In total, 334 of the species in our dataset (74%) migrate. We predict that the use of migration will require high efficiency for long cruising flights, and should therefore select for high aspect ratio. If speed is a factor on

migration, then wing loading should also be high; but if energy consumption is more important than speed, then wing loading should be low.

(8) *Commuting to feed young*

Foraging to feed young at the nest requires parent birds to commute between the nest and the feeding site. Furthermore, the food for the young must be carried back to the nest on the wing. This category includes all of the species in our dataset with altricial young, with the exception of the common cuckoo *Cuculus canorus*. In total, 382 of our species (85%) carry food to the nest to feed young. We predict that commuting to feed young will require both large flight forces and high flight efficiency. Commuting should therefore lead to high aspect ratio and low wing loading.

Because we score the ecological or behavioural characters as a '1' if the handbooks report that character as occurring at all in a given species, we are not making any subjective judgement about the importance of these characters to any particular species. This approach means that we only expect to detect associations between flight morphology and flight ecology where the effects are clear and strong: the first-order effects, rather than the subtleties, of adaptation.

7.4 Flight morphology

The relationship between the morphological data (see Chapter 5) and the ecological characters is represented in Figures 7.2 and 7.3, treating species as data points for the time being. Three things are immediately obvious from Figure 7.2. First, there is a clear relationship between body mass and aspect ratio, even though aspect ratio is dimensionless: larger birds have relatively longer and narrower wings than smaller birds. Second, the allometric relationship between aspect ratio and body mass is shallow, and appears to be strongly affected by a group of species with high aspect ratio and large body mass, which feed underwater and soar over the sea. Third, there are strong associations between some of our ecological characters and aspect ratio, at least when species are treated as data points. Similarly, Figure 7.3 shows a strong relationship between body mass and wing area—as is to be expected on scaling grounds—and there is a clear signal associated with most of the ecological characters. In fact, there seem to be strong associations between the morphological and ecological data for all of the ecological characters except migration and commuting. Although these associations are apparent when species are treated as data points, they obviously require quantitative statistical confirmation controlling for phylogeny (see Chapter 5).

7.5 Analyses

The predictions in the previous sections are of conditional form. For example, if a species is known to swim to catch prey, then we predict that its wing area will be smaller than if it were known not to swim to catch prey, other things being equal. Even

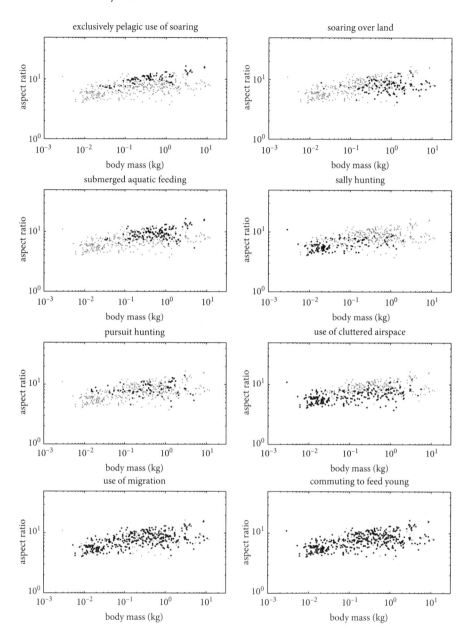

Figure 7.2 Plots of aspect ratio against body mass for each of the eight ecological predictors. Black points denote species that score '1' for a given character; grey points denote species that score '0'. The ecological characters appear to be associated with differences in aspect ratio and body mass in several cases, at least when species are treated as independent data points.

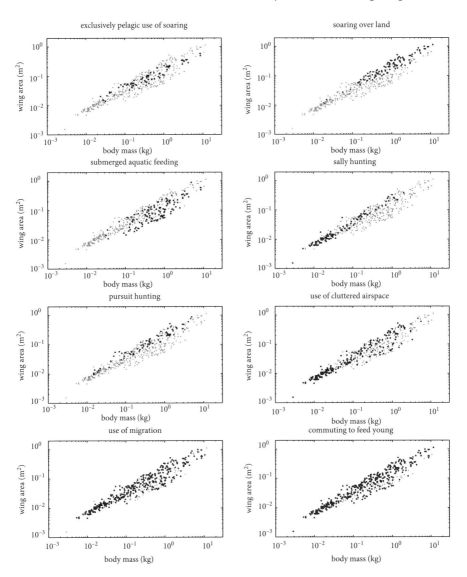

Figure 7.3 Plots of wing area against body mass for each of the eight ecological predictors. Black points denote species that score '1' for a given character; grey points denote species that score '0'. The ecological characters appear to be associated with differences in wing area and body mass in several cases, at least when species are treated as independent data points.

such obviously distinct traits as 'use of submerged aquatic feeding' or 'exclusively pelagic use of soaring', are not the result of atomized selection. The design of a Manx shearwater *Puffinus puffinus*, for example, is presumably a compromise between having wings that are of small enough area to be effective in swimming after prey, yet of sufficiently high aspect ratio to be effective in dynamic soaring. We therefore consider all of the ecological traits that we have identified together, rather than in isolation. Finally, because we are comparing across species, we will need to control for the specific error covariance structure that is expected to arise as a result of phylogenetic relatedness. Phylogenetically controlled generalized least squares estimates the expected value of a continuous response variable given a set of continuous and/or discrete predictor variables, while controlling for the confounding effects of phylogeny (Chapter 5). It is therefore ideally suited to testing predictions of the conditional form that we have made here.

A typical model in our analysis would fit log wing area, say, as a linear function of log body mass and the eight binary ecological predictors. After back-transformation, this amounts to fitting wing area as a power-law function of body mass, multiplied by a series of eight factors modelling the effects of each of the binary ecological predictors. We fit linear models of the same form for log aspect ratio, log wingspan and log wing chord. Each of these models requires 10 significance tests. We also fit a linear model for log body mass as a function of the eight ecological predictors, which requires a further nine significance tests. Given the multiplicity of tests, we only report results as being statistically significant after controlling the overall false discovery rate at the 5% level (Benjamini and Hochberg, 1995). We also record whether the same results would still have been significant after applying the more conservative Bonferroni correction to control the overall type I error rate for this chapter at the 5% level. We report the fitted scaling exponents with respect to body mass but emphasize that they are only being used to control for scale effects here, and should not be interpreted as estimates of a bivariate relationship between wing morphology and body mass (see also Chapter 4).

7.6 Results

Table 7.1 presents the results of the phylogenetically controlled generalized least squares analysis fitting log body mass as a linear function of the eight binary ecological predictors simultaneously. Three ecological characters are significant predictors of body mass. First, knowing that a bird soars over land allows us to predict it will have approximately twice the mass, on average, of a bird that does not soar over the land, other things being equal. Second, knowing that a bird hunts by sallying from perches allows us to predict that it will have a 25% lower body mass on average. Third, knowing that a bird has to contend with aerial clutter also allows us to predict that it will have a 28% lower body mass on average. None of the other ecological predictors is significantly associated with body mass in our phylogenetically controlled analysis. In the remainder of the analyses, we control for body mass when fitting the models.

Table 7.1 Results of the phylogenetically controlled generalized least squares analysis fitting log body mass as a linear function of the eight binary ecological predictors simultaneously. This analysis tests whether and how changes in the ecological characters are associated with changes in body mass. The p-values are two tailed and are based upon F-ratios calculated using type III adjusted sums of squares. We write $p \ll 0.01$ where $p \leq 0.0010$. Controlling the overall type I error rate for this chapter at the 5% level using a Bonferroni correction leads to a critical p-value of $0.05/49 = 0.0010$. It follows that any predictor for which $p \ll 0.01$ will be statistically significant even after Bonferroni correction. The fitted coefficients of the ecological predictors are shown in the penultimate column if the p-value for the predictor is statistically significant under the less conservative approach of controlling the overall false discovery rate for this chapter at the 5% level. These fitted coefficients combine additively with respect to log body mass, and therefore have a multiplicative effect upon body mass itself. The last column shows the percentage effect upon body mass of each statistically significant predictor.

Dependent variable:	log body mass		
Variation explained:	$R^2 = 0.09$		
Predictor	p-value	Coefficient	% effect
Intercept	$\ll 0.01$	−1.32	
Exclusively pelagic use of soaring	0.05		
Soaring over land	$\ll 0.01$	0.69	+99%
Submerged aquatic feeding	0.98		
Sally hunting	< 0.01	−0.33	−28%
Pursuit hunting	0.65		
Use of cluttered airspace	< 0.01	−0.28	−25%
Use of migration	0.70		
Commuting to feed young	0.15		

Hence, where we later find a statistically significant effect of soaring over land on wing area, this effect will be over and above any increase in wing area that is attributable to the scaling of wing area with body mass. In other words, although birds that soar over land certainly tend to have larger wings because they are more massive, this is not the whole story.

For a bird of a given body mass, our aerodynamic predictions relate principally to changes in wing area and aspect ratio. In Chapter 5 we reported that wing area scales strongly with body mass with an exponent of 0.668, which is statistically indistinguishable from the exponent of 0.667 expected under isometry. When the ecological predictors are included in the analysis, the scaling exponent becomes 0.654, with a 95% confidence interval (0.628, 0.679), which includes the exponent of 0.667 expected under isometry. Hence, although the model for log wing area explains 87% of the total variation (see Table 7.2), most of that explanatory power is due to the strong scaling of wing area with body mass. On the other hand, there is no simple reason

Table 7.2 Results of the phylogenetically controlled generalized least squares analysis fitting log wing area and log aspect ratio as linear functions of log body mass and the eight binary ecological predictors simultaneously. These analyses test whether and how changes in the ecological characters are associated with changes in wing area or aspect ratio, relative to their expectations for a bird of given body mass. We use the same reporting conventions as in Table 7.1.

Dependent variable:	log wing area			log aspect ratio		
Variation explained:	$R^2 = 0.87$			$R^2 = 0.10$		
Predictor	p-value	Coeff	% effect	p-value	Coeff	% effect
Intercept	≪ 0.01	−2.09		≪ 0.01	2.14	
Log body mass	≪ 0.01	0.65		≪ 0.01	0.03	
Exclusively pelagic use of soaring	≪ 0.01	0.25	+29%	< 0.01	0.12	+13%
Soaring over land	≪ 0.01	0.25	+29%	0.01	0.09	+9%
Submerged aquatic feeding	≪ 0.01	−0.23	−21%	0.02		
Sally hunting	0.98			0.59		
Pursuit hunting	0.10			0.19		
Use of cluttered airspace	0.46			0.23		
Use of migration	0.50			0.46		
Commuting to feed young	0.82			0.03		

coeff = coefficient.

to expect aspect ratio to scale strongly with body mass, because aspect ratio is dimensionless. Consequently, the model for aspect ratio against body mass and the eight ecological predictors explains only 10% of the variation in aspect ratio (Table 7.2). Nevertheless, aspect ratio still scales weakly with body mass, with an exponent of 0.034. The 95% confidence interval on the scaling exponent is (0.016, 0.053), which indicates that aspect ratio scales positively with respect to body mass. This means that larger birds have wings with a slightly longer, narrower planform than smaller birds, even after controlling for phylogeny and ecology. This confirms the positive scaling of aspect ratio with body mass that was previously demonstrated by Alerstam et al. (2007) on a smaller dataset using a phylogenetically controlled analysis which did not account for the effects of ecology.

Of the eight ecological characters, three are significant predictors of wing area, and two are significant predictors of aspect ratio (Table 7.2). Knowing that a bird soars over the sea allows us to predict that its wing area will be 29% larger, and that its aspect ratio will be 13% higher, all else considered. Likewise, knowing that a bird soars over the land allows us to predict it will have 29% larger wing area and 9% higher aspect ratio, other things considered. On the other hand, knowing that a bird submerges fully to feed allows us to predict only that it has 21% smaller wing area, all else considered. None of the other ecological characters is significantly correlated with either aspect ratio or wing area in our analyses. Because the models in Table 7.2

Table 7.3 Results of the phylogenetically controlled generalized least squares analysis fitting log wingspan and log wing chord as linear functions of log body mass and the eight binary ecological predictors simultaneously. These analyses test whether and how changes in the ecological characters are associated with changes in wingspan or wing chord, relative to their expectations for a bird of given body mass. We use the same reporting conventions as in Table 7.1.

Dependent variable:	log wingspan			log wing chord		
Variation explained:	$R^2 = 0.86$			$R^2 = 0.78$		
Predictor	p-value	Coeff	% effect	p-value	Coeff	% effect
Intercept	≪ 0.01	0.02		≪ 0.01	−2.11	
Log body mass	≪ 0.01	0.34		≪ 0.01	0.31	
Exclusively pelagic use of soaring	≪ 0.01	0.17	+19%	0.09		
Soaring over land	≪ 0.01	0.16	+17%	< 0.01	0.09	+10%
Submerged aquatic feeding	≪ 0.01	−0.08	−8%	≪ 0.01	−0.16	−14%
Sally hunting	0.77			0.89		
Pursuit hunting	0.10			0.63		
Use of cluttered airspace	0.13			0.13		
Use of migration	0.24			0.91		
Commuting to feed young	0.15			0.13		

coeff = coefficient.

control for body mass, each of these differences in wing area or aspect ratio relates to the difference that we would expect to see in a bird of given body mass according to whether or not it displayed the ecological character in question. Wing loading varies inversely with wing area, so increasing wing area at a given body mass causes a reduction in wing loading. It follows that birds that soar have reduced wing loading, whereas birds that submerge fully to feed have increased wing loading.

These changes in wing area and aspect ratio could in principle be produced by changes in wingspan and/or wing chord. Table 7.3 presents the results of the linear models that we fitted for log wingspan and log wing chord. These analyses are obviously not independent[5] of the analyses of wing area and aspect ratio that we presented in Table 7.2, but they show that almost all the available morphological options have been used as birds' wings have evolved in response to ecological and behavioural selection pressures. Unsurprisingly, the same three ecological characters that were significant predictors of wing area or aspect ratio are also significant predictors of wingspan and chord. For example, birds that submerge fully to feed have 8% smaller span and 14% smaller chord. Birds that soar exclusively over the sea have 19% higher span, but do not have significantly different wing chord, all else considered

---

[5] Strictly speaking, this non-independence violates the assumptions used to compute the false discovery rate, but the consequences of this are not severe.

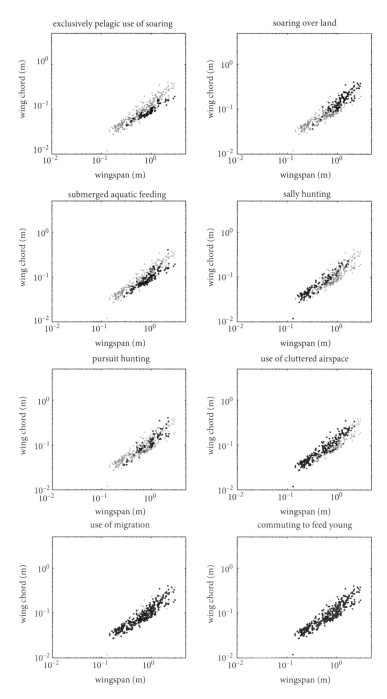

Figure 7.4 Plots of wing chord against wingspan for each of the eight ecological predictors. Black points denote species that score '1' for a given character; grey points denote species that score '0'. The plots have equal axes to make them more directly comparable with Figures 7.1 and 7.5.

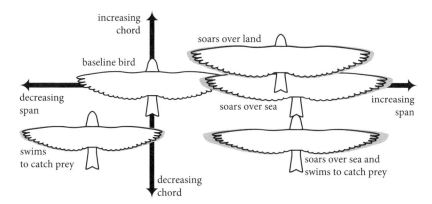

Figure 7.5 Statistically significant changes in flight morphology in response to ecological selection pressures, from the analyses for span and chord. All drawings are scaled relative to a baseline bird that represents the average wing morphology for a 0.1 kg bird. The drawings are scaled by the observed percentage changes in chord and span, with grey showing the extent of the 95% confidence intervals. The positions of the drawings have been shifted horizontally by the change in span and vertically by the change in chord.

(Table 7.3). On the other hand, birds that soar over the land have 17% higher span, but also have 10% higher chord. Hence, although both groups of soaring bird have significantly larger wing area and significantly higher aspect ratio than birds that do not soar, these changes in wing area and aspect ratio are accomplished in different ways. In birds that soar over the land, there is evidence that both wingspan and wing chord are increased, but that wingspan increases to a greater extent than wing chord. Conversely, in birds that soar exclusively over the sea, there is only statistical support for an increase in span. In each case, the result is an increase in both wing area and aspect ratio. We explore the differences between representatives of these two groups further in Chapter 8.

Figure 7.4 shows the relationship between span and chord for our 450 species, for comparison with the predictions presented in Figure 7.1. The scaling exponents of wing span and chord with body mass are 0.34 for span and 0.31 for chord, which combine to produce the overall scaling exponent of 0.03 that we found for aspect ratio, and hence the relatively longer and narrower wings of larger birds. The significant results from the analyses for span and chord are illustrated diagrammatically in Figure 7.5.

7.7 Conclusions

Body size in birds is strongly predicted by whether a species soars over the land, flies in clutter, or hunts by sallying. Although this conclusion relates directly to body mass, it has consequences for wing area and aspect ratio, which both scale with body mass.

There are also strongly significant effects on wing area and aspect ratio controlling for body mass in birds that soar, and in birds that submerge fully to feed. These changes in wing morphology result from changes in wingspan (soaring exclusively over the sea), or changes in both span and chord (soaring over land; submerging fully to feed). These associations are consistent with what we would expect to see as a result of selection for flight performance. The increased span of birds that soar over the sea is consistent with selection for straight glide performance. The increased span and chord in birds that soar over the land, is consistent with selection for cross-country soaring with its requirement for tight circling at low sink rate in thermal updrafts and efficient glide between thermals. We explore these effects further in Chapter 8. The decreased span and chord of birds that feed underwater is consistent with the use of the wings in a medium that is 1000 times denser than air. Perhaps as interesting are those ecological characters that have no significant effect on flight morphology. After all, we picked ecological and behavioural characters that we expected would have obvious direct links to flight performance. We found no significant effects on flight morphology of pursuit hunting, commuting to feed young, or migration.

The changes in flight morphology that we detected in our phylogenetically controlled analyses of birds are shown in Figure 7.5. The comparative approach that we have taken is a strong filter, which is only likely to detect strong signals through the noise expected in any large comparative database of morphological measurements. This is exacerbated by the relatively blunt process of scoring ecological characters recorded by field ornithologists in a binary form. Most importantly, our ecological characters are permissive: a score of '1' means only that a bird's flight morphology permits it to exhibit that character; it does not say anything about the importance of that character to the species. The statistically significant changes in flight morphology that are plotted in Figure 7.5 are therefore only those strongest of results that have made it through this filter. The plots of species as data points in Figures 7.2 to 7.4 suggest that other effects might appear in an ordinary least squares analysis without phylogenetic control, but there is no statistical support for these effects in our analyses with phylogenetic control. Hence, there is no statistical support in our data for a number of the relationships between flight morphology and flight ecology that might appear to exist when species are incorrectly treated as independent data points.

8

Trade-offs: selection, phylogeny, and constraint

8.1 Introduction

The comparative analyses of Chapter 7 showed that soaring birds have significantly higher aspect ratio wings and significantly lower wing loading than species that do not soar. This qualitative conclusion is consistent with the directional predictions that we made in Chapter 6, where we used dimensional analysis, supported by the results of classical wing theory, to establish how flight morphology is expected to respond to selection for different aspects of flight performance. Directional predictions of this sort are simple to formulate and straightforward to test, but they shed less light than we might hope on the trade-offs that exist between conflicting performance objectives. What we gain in simplicity, we lose in specificity. For example, the prediction that soaring birds should have higher aspect ratio wings than non-soaring birds holds for any kind of soaring flight (see Chapters 6 and 7). Nevertheless, it is evident that there is marked systematic variation in aspect ratio across those species which do specialize in soaring, and that this variation in aspect ratio is associated with variation in other morphological variables such as wing loading.

For example, petrels and albatrosses (Procellariiformes), which specialize in dynamic soaring over the oceans, have higher aspect ratio wings and higher wing loading given their body mass than the broad-winged raptors (Accipitriformes), which specialize in thermal soaring over the land. The discussion of phylogenetic non-independence in Chapter 5 should serve as a caution against blindly assuming that this systematic variation is necessarily the result of natural selection for different kinds of soaring. However, it is still reasonable to ask whether the detailed variation in flight morphology that we see is consistent with the hypothesis that species in these two taxa have been optimized for different kinds of soaring. In tackling this question, we will need to compare different aspects of flight morphology and different aspects of flight performance simultaneously, which we will do by making use of concepts drawn from the wider literature on multi-objective optimization. Our aim in this chapter is to use these concepts to develop a general framework for asking questions about performance trade-offs and multi-objective optimization in biology. We begin by briefly revisiting the theoretical basis for assuming that natural selection has an optimizing tendency, as developed in Chapter 2.

Evolutionary Biomechanics. Graham Taylor & Adrian Thomas.
© Graham Taylor & Adrian Thomas 2014. Published 2014 by Oxford University Press.

8.2 The adaptive landscape revisited

Chapter 2 presented a careful reformulation of Wright's adaptive landscape metaphor, which explains the exact sense in which natural selection can be expected to lead to phenotypic optimization. This view is fundamental to everything that follows in this chapter, so we begin by restating the case briefly in order to motivate our subsequent discussion of multi-objective optimization. In our 'drowning landscape' model of evolution, the 'vertical' axis of the landscape represents individual fitness relative to the population mean, whilst the 'horizontal' axes represent performance objectives for natural selection. We define a performance objective as any quantity whose increase would be expected to enhance the selective advantage of an allele conferring that increase, in the hypothetical case that such an increase could be effected without impacting performance in any other dimension (see Chapter 2). Implicit in this definition is the understanding that it is not usually possible in practice to improve performance in one dimension without degrading performance in another. The detailed form of the resulting trade-offs is jointly determined by the fitness function and by the constraints that relate the various performance objectives.

The process of natural selection itself is characterized by a directional flux in the intrinsic selective advantage of the alleles, which is expected to cause an increasing proportion of the population to sit within the higher occupied regions of the landscape at any moment in time. As adaptation spreads, the fitness of those individuals occupying the higher regions of the landscape is brought closer to the population mean ('sea level'), and the landscape itself slips beneath the waves. Those individuals that cling on at the waterline are no better off than their ancestors insofar as relative fitness is concerned, but if the environment remains reasonably stable, then they are likely to be better adapted to the prevailing conditions. Consequently, natural selection behaves like an optimization process, and will in fact appear to be maximizing individual fitness unless social effects or genetic conflict come in to play. This does not mean to say that a maximum of the fitness function will necessarily have been found, or indeed that a unique maximum exists—only that selection has a maximizing tendency.

We may formalize what we have just said mathematically by stating that natural selection behaves as if attempting to solve the following optimization problem:

$$\begin{aligned}
\text{maximize:} \quad & w(\boldsymbol{y}) \\
\text{subject to:} \quad & g_i(\boldsymbol{y}, \boldsymbol{x}) \geq 0 \quad \text{for } i = 1, \dots, k \\
& h_j(\boldsymbol{y}, \boldsymbol{x}) = 0 \quad \text{for } j = 1, \dots, m
\end{aligned}$$

in which $w(\boldsymbol{y})$ is the relative fitness function for the species in question, expressed in terms of the performance objectives contained in the vector $\boldsymbol{y}$ (see also Grafen, 2002). In the mathematical optimization literature, $w(\boldsymbol{y})$ would be called the objective function (e.g. Boyd and Vandenberghe, 2004). We assume that $\boldsymbol{x}$ is a vector of design variables sufficient to describe the phenotype of any individual completely, such that the k inequality constraints $g_i(\boldsymbol{y}, \boldsymbol{x})$ and m equality constraints $h_j(\boldsymbol{y}, \boldsymbol{x})$ include all of the constraints upon the problem—be they physical, informational, or developmental in nature. These constraints (see Chapter 3) are written in a standard form with zero on the right-hand side, so it is important to note that any equality or inequality can be put in this standard form by subtracting the right-hand side of the expression from both sides. There is therefore no magic in this formalization, which is merely a generic statement of a constrained optimization problem with $w(y)$ as its objective function. This problem statement is only useful if we can do something with it, which is the focus of the next section.

8.3 Multi-objective optimization

The problem statement in the previous section treats natural selection as attempting to solve a constrained scalar optimization problem, because although relative fitness $w(y)$ is a function of many variables, it is itself a scalar. In practice, the most that we can probably say about the relative fitness function is that some of the phenotypic performance objectives upon which it depends are expected to be weighted more strongly in some species than in others. Fortunately, the problem of having no prior knowledge of the detailed form of the objective function is not unique to biology. For instance, there are plenty of instances in engineering design where it pays not to be too dogmatic about the detailed form of the objective function, and a large literature has grown up about this topic under the name of multi-objective optimization or vector optimization (e.g. Collette and Siarry, 2003).

The basic idea of multi-objective optimization is to work with a set of key performance objectives upon which the overall objective function is known to depend; rather than working with the overall objective function, whose detailed form is unknown. Mathematically, a performance objective may be defined as any quantity with respect to which the partial derivative of the overall objective function is expected always to be positive. We earlier defined y as a vector of quantities whose increase would be expected to enhance the selective advantage of any allele conferring that increase, in the hypothetical case that such an increase could be effected without impacting performance in any other dimension. These two definitions amount to essentially the same thing if the interests of the individual and the alleles are aligned, and it follows that we may rewrite the earlier scalar optimization problem as a vector optimization problem:

$$\begin{aligned} \text{maximize:} \quad & y \\ \text{subject to:} \quad & g_i(y, x) \geq 0 \quad \text{for } i = 1, \ldots, k \\ & h_j(y, x) = 0 \quad \text{for } j = 1, \ldots, m. \end{aligned}$$

Because y is defined in relation to the selective advantage of the alleles, this vector optimization arguably has an even stronger claim to being an accurate description of natural selection than does the earlier scalar optimization in terms of the relative fitness of individuals. This is because natural selection is fundamentally characterized by the spread of alleles with positive intrinsic selective advantage, which only leads to the apparent maximization of individual fitness if the interests of the alleles happen to be aligned with the reproduction of their bearer (see Chapter 2).

Having rewritten the problem in this way, we are left with the statement that natural selection behaves as if maximizing all of the elements of y simultaneously, subject to the trade-offs implicit in the k equality constraints $g_i(y, x) = 0$ and m inequality constraints $h_j(y, x) \geq 0$. It follows that the alleles borne by a phenotype which performs as well as or better than another phenotype on every performance objective will always be favoured over the alleles borne by the other phenotype. In the terminology of multi-objective optimization, the first phenotype may therefore be said to dominate the second. In general, one phenotype may be said to dominate another if it performs as well or better on every performance objective.

A phenotype that solves the multi-objective optimization problem obviously cannot be dominated by any other phenotype that is present in the population, although there may be many other solutions that do not dominate it, and which it does not dominate. In the terminology of multi-objective optimization, these are called nondominated solutions, and the set of all non-dominated solutions is called the Pareto set (see e.g. Collette and Siarry, 2003; Boyd and Vandenberghe, 2004). The important point to note is that a phenotype cannot be a solution to the multi-objective optimization problem unless it is a member of the Pareto set, and this holds true regardless of how the various performance objectives are weighted in the relative fitness function. This simple result proves to be exceedingly useful if the constraints upon the problem are known, as is often the case in biomechanics (see Chapter 3). We illustrate this by example in the next section, with reference to the flight morphology of soaring birds.

8.4 Trade-offs in soaring flight

The statements about multi-objective optimization that we made in the previous section were made with regard to natural selection acting within a species. We cannot know in detail how natural selection weights different performance objectives in different species, and these weightings will certainly vary through time. We may, however, be able to make quite broad statements about the relative weighting of different performance objectives. This was the basis of the simple directional predictions that we made in Chapter 6, and is also the basis of the logic that underpins this chapter.

8.4.1 Performance objectives in soaring flight

Soaring flight is an excellent example to consider, because the phenomenon is both simple enough to be modelled given knowledge of only a few morphological variables, and complex enough that the optimal trade-offs among conflicting performance objectives vary markedly between environments. The essence of soaring is to overcome the dissipation of kinetic energy due to aerodynamic drag, by harvesting kinetic energy from the atmosphere and converting any surplus to potential energy.

This is only feasible for an animal with high aerodynamic efficiency, which manifests itself in a shallow glide angle, and always entails having a reasonably high-aspect ratio wing (Chapter 6). It is reasonable to assume that high aerodynamic efficiency will be key for any soaring animal, but this cannot be all that there is to the problem, because otherwise all soaring birds would look like albatrosses (Pennycuick, 1971).

In fact, having a shallow glide angle is only really an end in itself when it is import-ant to maximize the range attained in descent from a given altitude. It is difficult to think of cases where this is actually what matters to a soaring bird—except, perhaps in thermal-soaring species that cross channels between land masses on migration. In most cases, what matters most is: 1) to gain height quickly in sources of lift, and avoid losing too much height elsewhere; and 2) to cover ground quickly when travelling between sources of lift, or when travelling to a target. We may usefully think of these as mission requirements, in contradistinction to the performance objectives which fa-cilitate them. The first mission requirement—to gain height quickly in sources of lift, and avoid losing too much height elsewhere—hinges upon having a low sink rate. In species which soar over land, it also depends upon the ability to turn steadily in pos-sibly small thermals. The second mission requirement—to cover ground quickly—hinges upon having a high glide speed at minimum glide angle. The groundspeed at minimum glide angle is usually referred to simply as the best glide speed.

We may therefore identify three key performance objectives for soaring flight: 1) high best glide speed; 2) low minimum sink rate; and 3) low limiting turn radius. It is possible to estimate how a given species of bird scores on each of these perform-ance objectives from measurements of its aspect ratio ($\mathcal{R}$), wing area (S), and body mass (m). These three performance objectives are defined mathematically in Box 8.1, which details all of the necessary assumptions. An informal explanation now follows:

Box 8.1 Definitions of the three soaring performance objectives.

We collect here all of the mathematical identities used to solve for the three soaring per-formance objectives discussed in the main text, given measurements of aspect ratio ($\mathcal{R}$), wing area (S) and body mass (m). The notation used is the same as in Chapter 6, from which most of the identities are drawn. We make the same small angle approximations for glide angle as in Chapter 6, and assume elliptic loading.

(1) **Best glide speed** (U^*) is defined as the glide speed U that solves:

$$\text{maximize:} \quad \frac{C_L}{C_D} \qquad \text{(lift-to-drag ratio)}$$

$$\text{subject to:} \quad C_L = \frac{2mg}{\rho U^2 S} \qquad \text{(equilibrium lift coefficient)}$$

$$C_D = 0.01 + C_{Di} + C_{Df} \qquad \text{(total drag coefficient)}$$

$$C_{Di} = \frac{C_L{}^2}{\pi \mathcal{R}} \qquad \text{(induced drag coefficient)}$$

continued overleaf

Box 8.1 *Continued*

$$C_{Df} = 2\left(\frac{1.328}{\sqrt{Re}}\right) \qquad \text{(friction drag coefficient)}$$

$$Re = \frac{\rho U}{\mu}\sqrt{\frac{S}{\mathcal{R}}} \qquad \text{(chord Reynolds number)}$$

The formula for the friction drag coefficient is the classical laminar boundary-layer solution for a flat plate (Van Dyke, 1964). The constant 0.01 is the assumed body drag coefficient, referenced to wing area. There is some controversy in the literature over the appropriate value of this parameter, which is notoriously difficult to measure. The best field estimates of body drag are for passerines and give a mean body drag coefficient of 0.018 referenced to wing area (Hedenström and Liechti, 2001). The body drag coefficient of larger, streamlined soaring species may be expected to be somewhat smaller, so rounding down to 0.01 seems reasonable.

(2) **Minimum sink rate** (U_s^*) is defined as the solution of the minimization:

$$\text{minimize: } U_s = \sqrt{\frac{2mgC_D^2}{\rho S C_L^3}} \qquad \text{(sink rate)}$$

subject to all of the constraints just listed.

(3) **Limiting turn radius** (r^*) is defined as the solution of the minimization:

$$\text{minimize: } r = \frac{2m}{\rho S C_L \sin\phi} \qquad \text{(turn radius)}$$

$$\text{subject to: } C_L = \frac{2\pi\alpha}{1 + 2/\mathcal{R} + 16(\log(\pi\mathcal{R}) - 9/8)/(\pi\mathcal{R})^2} \qquad \text{(lift coefficient)}$$

$$0 \le \alpha \le \frac{\pi}{12} \qquad \text{(angle of attack)}$$

$$0 \le \phi \le \frac{\pi}{2} \qquad \text{(bank angle)}$$

where ϕ is the bank angle. The formula for the lift coefficient is the higher-order lifting-line approximation from Box 6.1. It is obvious by inspection that the solution of this minimization is found at maximum bank angle ($\phi = 90°$), and at maximum angle of attack ($\alpha = 15°$), which we assume to be the angle at which the wing stalls.

(1) Best glide speed (U^*) is defined as the glide speed at minimum glide angle, and is achieved at maximum lift-to-drag ratio (Box 8.1). We have shown already that the lift-induced component of the drag coefficient depends inversely upon aspect ratio and quadratically upon lift coefficient (Chapter 6). Lift must balance body weight at equilibrium, so for a given wing loading, flying at a lower airspeed will entail operating at a higher lift coefficient and therefore at a higher induced drag coefficient. Other components of the total drag coefficient do not show the same airspeed dependence, although there is a subtle Reynolds number-dependent effect of airspeed upon friction drag (Box 8.1). However, the total drag

force is proportional to the drag coefficient times the square of the airspeed, so the maximum lift-to-drag ratio is always achieved at an intermediate glide speed that appropriately balances these opposing effects. This optimum can be found numerically for each species and used to compute the best glide speed.

(2) In straight flight in still air, a bird's sink rate is equal to the sine of its glide angle multiplied by its airspeed. We have already shown that the airspeed of a gliding bird depends directly upon the square root of its wing loading and inversely upon the square root of its lift coefficient (Chapter 6). Birds can therefore fly at a lower airspeed by operating at a higher lift coefficient, but this will not improve their sink rate monotonically because of the increase in the induced drag coefficient that accompanies any increase in lift coefficient. Minimum sink rate (U_s^*) is therefore achieved at an airspeed only a little lower than the best glide speed (see Box 8.1).

(3) Soaring birds turn by banking so that the horizontal component of their lift vector provides the centripetal force needed to turn. This requires an increase in lift coefficient if the vertical component of the lift vector is still to support body weight. For a given bank angle, minimum turn radius is attained at maximum lift coefficient—and hence at the maximum angle of attack that the wings can reach without stalling. The tightest possible turn is achieved at an unrealistically high bank angle of 90°, and is equal to what Pennycuick (1971) called the 'limiting' turn radius (r^*). The minimum turn radius that can be achieved at any given bank angle is equal to the limiting turn radius divided by the sine of the bank angle (see also Norberg, 1990). It follows that the limiting turn radius actually limits performance in a meaningful way at all bank angles (Box 8.1). Other things being equal, a bird with a reduced limiting turn radius will be able to achieve the same given radius of turn by banking less steeply or by operating at a lower lift coefficient. This has obvious benefits for turning performance, because the alternative of operating at a very high lift coefficient or in a steeply banked turn inevitably results in a high sink rate. Low limiting turn radius is therefore an obvious performance objective for any bird that turns steadily whilst soaring.

The numerical values of our estimates of these three soaring performance objectives are strongly influenced by the numerical assumptions that we make about the body and friction drag coefficients (Box 8.1). However, the qualitative conclusions that we draw from them in the next section are robust to quite large variations in the assumed values of the numerical parameters. This is because we use the concept of Pareto optimality to compare the relative performance of different flight morphologies, which means that the absolute numerical values of the estimates are of secondary importance to their rank order.

8.4.2 Flight performance of soaring birds

Figure 8.1a plots wingspan against wing area and body mass for all of the soaring petrels and albatrosses (Procellariiformes) and all of the broad-winged raptors

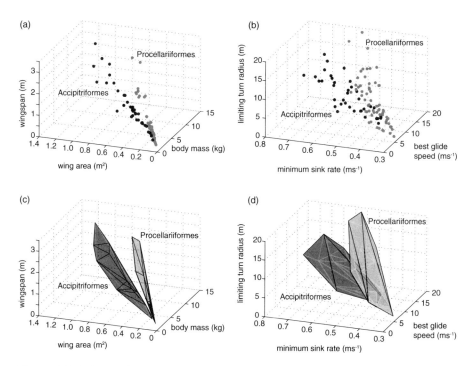

Figure 8.1 (a,c) Morphospace plot of wingspan against wing area and body mass for the 38 species of Accipitriformes (grey points) and 65 species of soaring Procellariiformes (black points) in our dataset. (b,d) Performance space plot of limiting turn radius against minimum sink rate and best glide speed for the same species. The volumes in (c,d) are the convex hulls containing all of the data points for each taxon in (a,b). The black lines on the surface of each convex hull join the data points forming the vertices of the nearside of the hull. The hulls are semi-transparent to allow the lines on the far side of the hulls also to be visualized. It is clear from this figure that the Accipitriformes and Procellariiformes occupy different regions of both the morphospace and the performance space.

(Accipitriformes) in our dataset. These taxa typify the two distinctive kinds of soaring flight morphology that Lilienthal (1889) first observed when comparing seabirds with soaring land-birds such as storks, cranes and raptors (see Chapter 7). It is already clear from Figure 8.1a that the two taxa occupy different regions within the morphospace that the three morphological variables define. Their occupancy of different regions of the morphospace is shown even more clearly by Figure 8.1c, which plots the convex hulls containing all of the points in each taxon. It is interesting to note that the occupied regions of the morphospace do not intersect at all for the 103 species plotted here, which represent approximately 18% of all Accipitriformes and 52% of all Procellariiformes.

Figure 8.1c has obvious resonance with Raup's classic analyses of shell coiling in molluscs (Raup, 1966; see also McGhee, 2007; Pigliucci, 2012). However, because we are able to map morphology directly onto performance, we have also provided two further graphs that plot the same data in the performance space defined by the three soaring performance objectives (Figure 8.1b,d). This shows explicitly how occupancy of different regions of the morphospace leads to occupancy of different regions of the performance space. The distinction between our new performance space and the classical morphospace (McGhee, 2007) parallels our earlier redefinition of the 'horizontal' axes of the adaptive landscape in terms of performance objectives rather than morphological variables. Crucially, the occupancy of different regions of the performance space by the Procellariiformes and Accipitriformes is precisely what we would expect to see if natural selection had weighted the three conflicting performance objectives differently in the two taxa. This is because different regions of the performance space effectively represent different resolutions of the trade-offs that exist between the three conflicting performance objectives.

Plotting the performance objectives in pairs clarifies the precise nature of the alternative resolutions that have been reached in these two taxa (Figure 8.2). In particular, the Accipitriformes are most clearly distinguished from the Procellariiformes by their lower limiting turn radius for a given best glide speed (Figure 8.2a). This is exactly what we would expect from first principles, given that thermal-soaring species must be able to turn tightly in order to exploit small thermals. In fact, raptors spend most of their flight time circling, so it is not surprising that their flight morphology is consistent with optimization for steady turning flight performance. In contrast, Procellariiformes perform better than Accipitriformes on the two

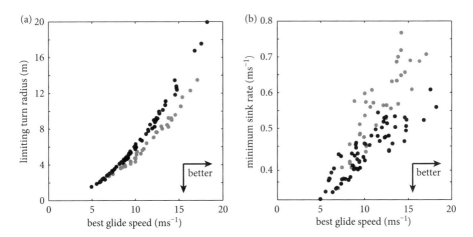

Figure 8.2 Scores on pairs of soaring performance objectives for the 38 species of Accipitriformes (grey points) and 65 species of soaring Procellariiformes (black points) in our dataset. (a) Accipitriformes tend to have a better limiting turn radius than Procellariiformes for a given best glide speed. (b) Procellariiformes tend to have a better minimum sink rate than Accipitriformes for a given best glide speed.

performance objectives associated with straight flight performance. Specifically, Procellariiformes typically have a lower minimum sink rate for a given best glide speed than Accipitriformes (Figure 8.2A), which is a reflection of the fact that their higher aspect ratio wings confer a better glide angle, and hence a better sink rate for a given airspeed. One interpretation of this is that the soaring flight morphology of Procellariiformes is less strongly compromised by the demands of steady turning than is the flight morphology of Accipitriformes.

8.4.3 Pareto optimality in soaring birds

In concluding our discussion of multi-objective optimization in soaring flight, we return now to the concept of Pareto optimality that we introduced in Section 8.3. Recall that one phenotype may be said to dominate another in respect of a set of performance objectives if, and only if, the first phenotype performs as well as or better than the second phenotype on every performance objective. Mathematically, we may write that a phenotype whose score on a given set of performance objectives y is denoted y_1 dominates another phenotype with scores denoted by y_2 if and only if $y_1 \geq y_2$. The subset of phenotypes that is not dominated by any other is called the Pareto set, and each of the phenotypes that this set contains may be said to be Pareto-optimal with respect to the specified set of performance objectives. Within the Pareto set, there will usually be some phenotypes which perform better than others on one or more performance objectives. However, by the definition just given, no phenotype within the Pareto set will equal or outperform any other phenotype within the Pareto set on every one of the performance objectives.

It is important to note that the concept of a Pareto-optimal phenotype differs markedly from the usual concept of optimality that is familiar in behavioural ecology and biomechanics. The concept of a Pareto-optimal phenotype is, in fact, much closer to the concept of an evolutionarily stable strategy that is familiar from evolutionary game theory. We refrain from making any more precise connection here because the concept of Pareto optimality is used in a very particular way in the mathematical theory of games. It is important to note that the claim that a phenotype is Pareto-optimal is weaker than the claim that a phenotype is optimal in the usual sense of being a local maximum or minimum of some function. Rather than claiming that a phenotype actually optimizes any one performance objective, which is what classical optimal foraging theory aims to test (but see Sober, 2008), we merely claim that there is no existing phenotype which is unequivocally better over a specified set of performance objectives. In principle, this means that any one member of the Pareto set could be optimal in the usual sense of the word, depending upon the relative importance of each of the conflicting performance objectives.

When all three soaring performance objectives are considered, the Pareto set identifies 87 of the 103 species (i.e. 84% of species) of Accipitriformes and soaring Procellariiformes in our dataset. In principle, this means that if overall soaring performance were assumed to be a weighted sum of the three performance objectives,

then it would be possible to find a particular set of weights for the different performance objectives that would make any of the 87 Pareto-optimal phenotypes optimal in
the usual sense of the word. That is to say, most of the variation in wingspan, wing
area, and body mass that we see can in principle be explained by variation in the relative importance to each species of these three conflicting performance objectives.
We do not mean to imply that this is actually the case, but rather to point out the rich
variation that can result from even a simple tri-objective optimization if the weights
of the conflicting performance objectives are varied.

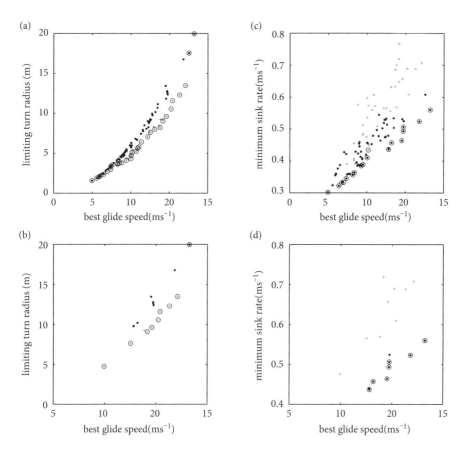

Figure 8.3 Pareto sets for two pairs of soaring performance objectives. (a,c) All 38 species
of Accipitriformes (grey points) and 65 species of soaring Procellariiformes (black points) in
our dataset. (b,d) Subset of all nine species of vulture (grey points: Cathartidae, Gypaetinae,
Aegyptiinae) and all nine species of albatrosses (black points: Diomedeidae) in our dataset. All
of the points contained in the Pareto set for each graph are circled. Note that Accipitriformes,
especially the vultures, tend to be Pareto-optimal in respect of limiting turn radius and best
glide speed (a,b). Procellariiformes, especially the albatrosses, tend to be Pareto-optimal in
respect of minimum sink rate and best glide speed (c,d).

It is of particular interest to plot the Pareto sets in respect of the pairs of performance objectives that best distinguish Accipitriformes from Procellariiformes, and vice versa (Figure 8.3a,c). Among the 38 species of Accipitriformes, 66% of species are Pareto-optimal in respect of best glide speed and limiting turn radius, but only 5% are Pareto-optimal in respect of best glide speed and minimum sink rate (Figure 8.3a). Conversely, 25% of the 65 species of Procellariiformes are Pareto-optimal in respect of best glide speed and minimum sink rate, but only 14% are Pareto-optimal in respect of best glide speed and limiting turn radius (Figure 8.3c). Thus, the Pareto set for best glide speed and limiting turn radius identifies an accipitriform rather than a procellariiform with odds of 4.8. This confirms that the flight morphology of Accipitriformes is better suited to steady turning flight performance than is the flight morphology of Procellariiformes. Conversely, the Pareto set for best glide speed and minimum sink rate identifies a procellariiform rather than an accipitriform with odds of 4.7. This confirms that the flight morphology of Procellariiformes is better suited to straight flight performance than is the flight morphology of Accipitriformes.

The conclusions that we have drawn so far relate to Procellariiformes and Accipitriformes as a whole; but if the reasoning that we have developed around them is correct, then we should expect the same conclusions to be even clearer in respect of the most specialized soaring species. These are the albatrosses (Procellariiformes: Diomedeidae), the New World vultures (Accipitriformes: Cathartidae) and the Old World vultures (Accipitriformes: Gypaetinae and Aegyptiinae). All nine species of vulture and all nine species of albatross are Pareto-optimal with respect to each other when measured against all three performance objectives. In contrast, the Pareto set for limiting turn radius and best glide speed includes eight of the nine vultures, but only one of the nine albatrosses (Figure 8.3b). Conversely, the Pareto set for minimum sink rate and best glide speed includes seven of the nine albatrosses but none of the vultures (Figure 8.3d). This confirms that the flight morphology of vultures is indeed better suited to steady turning flight, and that the flight morphology of albatrosses is better suited to straight flight. This makes good sense in light of the fact that vultures spend most of their flight time circling, whereas albatrosses spend most of their flight time gliding along a gently winding or curving course, interspersed with occasional fast—and presumably unsteady—turns (see Sachs et al., 2012).

It is worth commenting briefly upon why one species of albatross appears in the Pareto set for limiting turn radius and best glide speed (Figure 8.3b). This is because the Pareto set must always, by definition, include any species which is strictly the best among species for any one of the performance objectives being considered. The southern royal albatross *Diomedea epomophora* has the highest estimated best glide speed of any of the birds in our dataset, and is therefore a member of all of the Pareto sets plotted in Figure 8.3. Nothing specific should be read into the fact that one species of vulture does not appear in the Pareto set for limiting turn radius and best glide speed, and that two species of albatross do not appear in the Pareto set for minimum sink rate and best glide speed. A quick visual inspection of Figure 8.3c,d will be sufficient to convince that this is merely a consequence of thresholding that results from ranking the species in order to identify the Pareto set.

8.5 Conclusions

The theoretical framework that we have developed for analysing multi-objective optimization is applicable to any problem in which the constraints upon conflicting performance objectives can be identified. For example, the concept of Pareto optimality has been recently used and discussed in relation to the multi-objective optimization of gene networks and metabolic pathways (e.g. Schuetz et al., 2011; Warmflash et al., 2012; Pozo et al., 2012; Higuera et al., 2012). In a morphological context, Eloy (2013) has recently used an evolutionary algorithm to identify Pareto-optimal designs for undulatory swimming in a bi-objective optimization of energetic cost and swimming speed.

Shoval et al. (2012) have also used the concept of Pareto optimality to analyse trade-offs in several morphological examples, including aspect ratio and body mass in bats. However, they assumed a linear decrease in performance with distance from each of three 'archetypes' which they supposed would uniquely maximize performance in one dimension (cf. the complexity of Box 8.1). Shoval et al. (2012) claim to relax both assumptions later in their analysis, but they nevertheless reach the conclusion that phenotypes which are optimized with respect to three conflicting performance objectives will be distributed within an approximately triangular area of any relevant two-dimensional morphospace. It should be obvious, however, that there can be no meaningful distance metric in a morphospace that mixes a scale-invariant parameter such as aspect ratio with a scale-variant parameter such as body mass. As such, the analysis by Shoval et al. would appear to lack any theoretical foundation; its statistical flaws have been exhaustively discussed by Edelaar (Edelaar, 2013; but see Shoval et al., 2013).

In contrast, the empirical analysis that we have presented here is based upon a rigorous theoretical foundation (Chapter 2), and makes no assumptions about distance metrics. We have simply taken the existing interspecific variation in phenotype (Chapter 7), and used this to estimate the variation in performance in conflicting dimensions, based upon the theoretical physical constraints upon the problem (Chapter 6). This allows us to map points from the original morphospace into a new performance space, so as to test whether the phenotypes that we observe are consistent with the hypothesis that different performance objectives are weighted differently in different taxa. In the particular example of soaring flight morphology that we have considered here, the variation in flight morphology that we observe is consistent with the hypothesis that vultures are optimized for steady turning flight, whereas albatrosses are primarily optimized for straight flight. This is consistent with what we know about the ecology and behaviour of these taxa, and offers a convincing explanation of why vultures do not have efficient high-aspect ratio wings: their straight soaring flight performance is compromised by the fact that they spend most of their flight time circling.

It is difficult to imagine a clearer or simpler result than this to illustrate the power of an optimization analysis that takes account of multiple conflicting performance objectives simultaneously. The full grandeur of life on Earth is infinitely too complex and too varied to be captured by a few simple variables, of course, but the multitudinous trade-offs that characterize living organisms are always due ultimately to constraints of the same fundamental kind as we have identified here. The gradual diversification of species that so fascinated Darwin may be thought of as the process of finding alternative trade-offs among conflicting performance objectives in organisms subject to similar constraints.

True innovation is brought about when a change in either the genes or the environment introduces new performance objectives for natural selection, or fundamentally alters the constraints that relate them. When this happens, it is not merely the balance of trade-offs among conflicting performance objectives that changes, but the very rules of the game. Such game-changing innovations as the origin of walking from swimming, running from walking, or flight from falling, can all be thought of as functional or adaptive shifts (see also Simpson, 1944) that release new degrees of freedom—and with them new opportunities for selection. In the same way, we see a rich future for evolutionary biomechanics if the field can achieve a similar functional shift from being a niche specialism aimed at studying the evolution of biomechanical systems, to becoming a broader discipline aimed at understanding evolution through the study of biomechanics.

References

Alerstam, T., Rosén, M., Bäckman, J., Ericson, P. G. P. and Hellgren, O. (2007). Flight speeds among bird species: allometric and phylogenetic effects. *PLoS Biology* **5**, e197.

Alexander, R. M. (1976). Estimates of speeds of dinosaurs. *Nature* **261**, 129–130.

Alexander, R. M. (2003). *Principles of Animal Locomotion*. Princeton, N.J.: Princeton University Press.

Alexander, R. M. and Jayes, A. S. (1983). A dynamic similarity hypothesis for the gaits of quadrupedal mammals. *J. Zool., Lond.* **201**, 135–152.

Anderson, J. D. J. (1997). *A History of Aerodynamics and its Impact on Flying Machines*. Cambridge: Cambridge University Press.

Baker, G. L. and Blackburn, J. A. (2005). *The Pendulum: A Case Study in Physics*. Oxford: Oxford University Press.

Banavar, J. R., Damuth, J., Maritan, A. and Rinaldo, A. (2002). Supply-demand balance and metabolic scaling. *PNAS* **99**, 10506–10509.

Banavar, J. R., Maritan, A. and Rinaldo, A. (1999). Size and form in efficient transportation networks. *Nature* **399**, 130–132.

Barenblatt, G. I. (2003). *Scaling*. Cambridge: Cambridge University Press.

Batty, C. J. K., Crewe, P., Grafen, A. and Gratwick, R. (2013). Foundations of a mathematical theory of darwinism. *J. Math. Biol.* [doi:10.1007/s00285-013-0706-2].

Bels, V. L., Gasc, J. P. and Casinos, A., eds. (2003). *Vertebrate Biomechanics and Evolution*. Oxford: BIOS Scientific Publishers Ltd.

Benjamini, Y. and Hochberg, Y. (1995). Controlling the false discovery rate: a practical and powerful approach to multiple testing. *J. R. Stat. Soc. B* **57**, 289–300.

Bertram, J. E. A. (2004). New perspectives on brachiation mechanics. *Am. J. Phys. Anthropol.* **Suppl 39**, pp. 100–117.

Bertram, J. E. A. and Chang, Y.-H. (2001). Mechanical energy oscillations of two brachiation gaits: measurement and simulation. *Am. J. Phys. Anthropol.* **115**, 319–326.

Bertram, J. E. A., Ruina, A., Cannon, C. E., Chang, Y.-H. and Coleman, M. J. (1999). A point-mass model of gibbon locomotion. *J. Exp. Biol.* **202**, 2609–2617.

Biewener, A. A. (2003). *Animal Locomotion*. Oxford: Oxford University Press.

Bijma, P. (2009). Fisher's fundamental theorem of inclusive fitness and the change in fitness due to natural selection when conspecifics interact. *J. Evol. Biol.* **23**, 194–206.

Blickhan, R. (1989). The spring-mass model for running and hopping. *J. Biomech.* **22**, 1217–1227.

Blickhan, R. and Full, R. J. (1993). Similarity in multilegged locomotion: bouncing like a monopode. *J. Comp. Physiol. A* **173**, 509–517.

Boyd, S. P. and Vandenberghe, L. (2004). *Convex Optimization*. Cambridge: Cambridge University Press.

Brewer, M. L. and Hertel, F. (2007). Wing morphology and flight behavior of pelecaniform seabirds. *J. Morphol.* **268**, 866–877.

Bridgman, P. W. (1922). *Dimensional Analysis*. Yale: Yale University Press.

Bruderer, B. and Boldt, A. (2001). Flight characteristics of birds: I. radar measurements of speed. *Ibis* **143**, 178–204.

Bruderer, B., Peter, D., Boldt, A. and Liechti, F. (2010). Wing-beat characteristics of birds recorded with tracking radar and cine camera. *Ibis* **152**, 272–291.

Buckingham, E. (1914). On physically similar systems; illustrations of the use of dimensional equations. *Phys. Rev.* **4**, 345–376.

Bullimore, S. R. and Burn, J. F. (2006). Dynamically similar locomotion in horses. *J. Exp. Biol.* **209**, 455–465.

Bullimore, S. R. and Donelan, J. M. (2008). Criteria for dynamic similarity in bouncing gaits. *J. Theor. Biol.* **50**, 339–348.

Butler, M. A. and King, A. A. (2004). Phylogenetic comparative analysis: a modeling approach for adaptive evolution. *Am. Nat.* **64**, 683–695.

Butler, M. A., Schoener, T. W. and Losos, J. B. (2000). The relationship between sexual size dimorphism and habitat use in Greater Antilean *Anolis* lizards. *Evolution* **54**, 259–272.

Carroll, R. J. and Ruppert, D. (1996). The use and misuse of orthogonal regression in linear errors-in-variables models. *Am. Stat.* **50**, 1–6.

Cavagna, G. A., Heglund, N. C. and Taylor, C. R. (1977). Mechanical work in terrestrial locomotion: two basic mechanisms for minimizing energy expenditure. *Am. J. Physiol.* **233**, R243–R261.

Cavagna, G. A., Heglund, N. C. and Willems, P. A. (2005). Effect of an increase in gravity on the power output and the rebound of the body in human gravity. *J. Exp. Biol.* **208**, 2333–2346.

Chan, N. R., Dyke, G. J. and Benton, M. J. (2013). Primary feather lengths may not be important for inferring the flight styles of Mesozoic birds. *Lethaia* **46**, 146–153.

Chang, Y.-H., Bertram, J. E. A. and Lee, D. V. (2000). External forces and torques generated by the brachiating White-Handed Gibbon (*Hylobates lar*). *Am. J. Phys. Anthropol.* **113**, 201–216.

Cheng, C.-L. and Van Ness, J. W. (1999). *Statistical Regression with Measurement Error*, Kendall's Library of Statistics, 6. London: Arnold.

Clarke, A., Rothery, P. and Isaac, N. J. B. (2010). Scaling of basal metabolic rate with body mass and temperature in mammals. *J. Anim. Ecol.* **79**, 610–619.

Collette, Y. and Siarry, P. (2003). *Multiobjective Optimization: Principles and Case Studies*. Berlin: Springer-Verlag.

Cramp, S. (1977). *Handbook of the Birds of Europe, the Middle East, and North Africa: The Birds of the Western Palearctic. Volume 1: Ostrich to Ducks*. Oxford: Oxford University Press.

Cramp, S., ed. (1980). *Handbook of the Birds of Europe, the Middle East, and North Africa: The Birds of the Western Palearctic. Volume 2: Hawks to Bustards*. Oxford: Oxford University Press.

Cramp, S., ed. (1983). *Handbook of the Birds of Europe, the Middle East and North Africa: The Birds of the Western Palearctic. Volume 3: Waders to Gulls*. Oxford: Oxford University Press.

Cramp, S., ed. (1985). *Handbook of the Birds of Europe, the Middle East, and North Africa: The Birds of the Western Palearctic. Volume 4: Terns to Woodpeckers*. Oxford: Oxford University Press.

Cramp, S., ed. (1988). *Handbook of the Birds of Europe, the Middle East, and North Africa: The Birds of the Western Palearctic. Volume 5: Tyrant Flycatchers to Thrushes*. Oxford: Oxford University Press.

Cramp, S. and Brooks, D. J., eds. (1992). *Handbook of the Birds of Europe, the Middle East, and North Africa: The Birds of the Western Palearctic. Volume 6: Warblers.* Oxford: Oxford University Press.

Cramp, S. and Perrins, C. M., eds. (1994a). *Handbook of the Birds of Europe, the Middle East, and North Africa: The Birds of the Western Palearctic. Volume 8: Crows to Finches.* Oxford: Oxford University Press.

Cramp, S. and Perrins, C. M., eds. (1994b). *Handbook of the Birds of Europe, the Middle East, and North Africa: The Birds of the Western Palearctic. Volume 9: Buntings and New World Warblers.* Oxford: Oxford University Press.

Cramp, S., Perrins, C. M. and Brooks, D. J., eds. (1993). *Handbook of the Birds of Europe, the Middle East, and North Africa: The Birds of the Western Palearctic. Volume 7: Flycatchers to Shrikes.* Oxford: Oxford University Press.

Cresswell, W. (1994). Song as a pursuit-deterrent signal, and its occurrence relative to other anti-predation behaviours of skylark (*Alauda arvensis*) on attack by merlins (*Falco columbarius*). *Behav. Ecol. Sociobiol.* **34**, 217–223.

Darwin, C. R. (1842). *The Structure and Distribution of Coral Reefs. Being the First Part of the Geology of the Voyage of the Beagle, under the Command of Capt. Fitzroy, R.N. during the Years 1832 to 1836.* London: Smith Elder and Co.

Darwin, C. R. (1859). *On the Origin of Species by Means of Natural Selection, or the Preservation of Favoured Races in the Struggle for Life,* 1st Ed. London: John Murray.

Dawkins, R. (1986). *The Blind Watchmaker.* Harlow: Longman Scientific and Technical.

Dawkins, R. (1996). *Climbing Mount Improbable.* New York: Norton.

Dawson, T. H. (2001). Similitude in the cardiovascular system of mammals. *J. Exp. Biol.* **204**, 395–407.

DeJong, M. J. (1983). *Bounding Flight in Birds.* PhD thesis, University of Wisconsin, Madison, USA.

del Hoyo, J., Elliott, A. and Sargatal, J., eds. (1992). *Handbook of the Birds of the World. Volume 1: Ostrich to Ducks.* Barcelona: Lynx Edicions.

del Hoyo, J., Elliott, A. and Sargatal, J., eds. (1996). *Handbook of the Birds of the World. Volume 3: Hoatzin to Auks.* Barcelona: Lynx Edicions.

del Hoyo, J., Elliott, A. and Sargatal, J., eds. (1997). *Handbook of the Birds of the World. Volume 4: Sandgrouse to Cuckoos.* Barcelona: Lynx Edicions.

del Hoyo, J., Elliott, A. and Sargatal, J., eds. (2005). *Handbook of the Birds of the World. Volume 10: Cuckoo-shrikes to Thrushes.* Barcelona: Lynx Edicions.

del Hoyo, J., Elliott, A. and Sargatal, J., eds. (2006). *Handbook of the Birds of the World. Volume 11: Old World Flycatchers to Old World Warblers.* Barcelona: Lynx Edicions.

del Hoyo, J., Elliott, A. and Sargatal, J., eds. (2009). *Handbook of the Birds of the World. Volume 14: Bush-shrikes to Old World Sparrows.* Barcelona: Lynx Edicions.

Demoll, R. (1918). *Der Flug der Insekten und der Vögel: eine Gegenüberstellung.* Jena: G. Fischer.

Díaz-Uriarte, R. and Garland, T. J. (1996). Testing hypotheses of correlated evolution using phylogenetically independent contrasts: sensitivity to deviations from Brownian motion. *Syst. Biol.* **45** 27–47.

Díaz-Uriarte, R. and Garland, T. J. (1998). Effects of branch length errors on the performance of phylogenetically independent contrasts. *Syst. Biol.* **47**, 654–672.

Dietrich, M. R. and Skipper Jr., R. A. (2012). A shifting terrain: a brief history of the adaptive landscape. In: *The Adaptive Landscape in Evolutionary Biology* (ed. Svensson, E. I. and Calsbeek, R.), pp. 3–15. Oxford: Oxford University Press.

Donelan, J. M. and Kram, R. (2000). Exploring dynamic similarity in human running using simulated reduced gravity. *J. Exp. Biol.* **203**, 2405–2415.

Dudley, R. (2000). *The Biomechanics of Insect Flight : Form, Function, Evolution.* Princeton, NJ: Princeton University Press.

Duncan, R. P., Forsyth, D. M. and Hone, J. (2007). Testing the metabolic theory of ecology: allometric scaling exponents in mammals. *Ecology* **88**, 324–333.

Edelaar, P. (2013). Comment on "Evolutionary trade-offs, Pareto optimality, and the geometry of phenotype space". *Science* **339**, 757.

Edwards, A. W. F. (1994). The Fundamental Theorem of natural selection. *Biol. Rev.* **69**, 443–474.

Edwards, E. L. (1977). *The Story of the Pendulum Clock.* Altrincham: John Sherratt and Son.

Einstein, A. (1934). On the method of theoretical physics. *Phil. Sci.* **1**, 163–169.

Eloy, C. (2013). On the best design for undulatory swimming. *J. Fluid Mech.* **717**, 48–89.

Ennos, A. R. (2012). *Solid Biomechanics.* Princeton, NJ: Princeton University Press.

Enquist, B. J. and Niklas, K. J. (2002). Global allocation rules for patterns of biomass partitioning in seed plants. *Science* **295**, 1517–1520.

Etienne, R. S., Apol, M. E. F. and Olff, H. (2006). Demystifying the West, Brown and Enquist model of the allometry of metabolism. *Funct. Ecol.* **20**, 394–399.

Ewens, W. J. (1989). An interpretation and proof of the Fundamental Theorem of natural selection. *Theor. Pop. Biol.* **36**, 167–180.

Ewens, W. J. (1992). An optimizing principle of natural selection in evolutionary population genetics. *Theor. Pop. Biol.* **42**, 333–346.

Ewens, W. J. (2004). *Mathematical Population Genetics. I. Theoretical Introduction*, 2nd Ed. New York: Springer.

Ewens, W. J. (2011). What is the gene trying to do? *Brit. J. Phil. Sci.* **62**, 155–176.

Farley, C. T., Glasheen, J. and McMahon, T. A. (1993). Running springs: speed and animal size. *J. Exp. Biol.* **185**, 71–86.

Felsenstein, J. (1985). Phylogenies and the comparative method. *Am. Nat.* **125**, 1–15.

Ferguson-Lees, J. and Christie, D. A. (2001). *Raptors of the World.* London: Christopher Helm.

Fisher, R. A. (1930). *The Genetical Theory of Natural Selection*, 1st Ed. Oxford: Clarendon Press.

Fisher, R. A. (1941). Average excess and average effect of a gene substitution. *Ann. Eugen.* **11**, 53–63.

Fisher, R. A. (1958). *The Genetical Theory of Natural Selection*, 2nd Ed. New York: Dover.

Frank, S. A. (1997). The Price equation, Fisher's fundamental theorem, kin selection, and causal analysis. *Evolution* **51**, 1712–1729.

Frank, S. A. (2012). Wright's adaptive landscape versus Fisher's fundamental theorem. In: *The Adaptive Landscape in Evolutionary Biology* (ed. Svensson, E. I. and Calsbeek, R.), pp. 41–57. Oxford: Oxford University Press.

Freckleton, R. P. (2009). The seven deadly sins of comparative analysis. *J. Evol. Biol.* **22**, 1367–1375.

Freckleton, R. P., Cooper, N. and Jetz, W. (2011). Comparative methods as a statistical fix: the dangers of ignoring an evolutionary model. *Am. Nat.* **178**, E10–E17.

Galilei, G. (1638). *Discorsi e dimostrazioni matematiche intorno à due nuoue scienze attenenti alla mecanica i movimenti locali.* Leida: Appresso gli Elsevirii.

Garland, T. J. (2000). Using the past to predict the present: confidence intervals for regression equations in phylogenetic comparative methods. *Am. Nat.* **155**, 346–364.

Gehr, P., Mwangi, D. K., Ammann, A., Maloiy, G. M. O., Taylor, C. R. and Weibel, E. R. (1981). Design of the mammalian respiratory system. V. Scaling morphometric pulmonary diffusing capacity to body mass: wild and domestic mammals. *Resp. Physiol.* **44**, 61–86.

Geyer, H., Seyfarth, A. and Blickhan, R. (2006). Compliant leg behaviour explains basic dynamics of walking and running. *Proc. R. Soc. Lond. B.* **273**, 2861–2867.

Gillooly, J. F., Charnov, E. L., West, G. B., Savage, V. M. and Brown, J. H. (2002). Effects of size and temperature on developmental time. *Nature* **417**, 70–73.

Glazier, D. S. (2005). Beyond the '3/4-power law': variation in the intra- and interspecific scaling of metabolic rate in animals. *Biol. Rev.* **80**, 611–662.

Glazier, D. S. (2010). A unifying explanation for diverse metabolic scaling in animals and plants. *Biol. Rev.* **85**, 111–138.

Gould, S. J. (2002). *The Structure of Evolutionary Theory*. Cambridge, Massachusetts: Harvard University Press.

Grafen, A. (1989). The phylogenetic regression. *Phil. Trans. R. Soc. Lond. B* **326**, 119–157.

Grafen, A. (2002). A first formal link between the Price equation and an optimization program. *J. Theor. Biol.* **217**, 75–91.

Grafen, A. (2003). Fisher the evolutionary biologist. *J. Roy. Stat. Soc. D* **52**, 319–329.

Grafen, A. (2007). The formal Darwinism project: a mid-term report. *J. Evol. Biol.* **20**, 1243–1254.

Grafen, A. (2008). The simplest formal argument for fitness optimization. *J. Genet.* **87**, 421–433.

Greenwalt, C. H. (1962). *Dimensional Relationships for Flying Animals*, Smithsonian Miscellaneous Collections, 144. Washington: Smithsonian Institution.

Greenwalt, C. H. (1975). The flight of birds: the significant dimensions, their departure from the requirements for dimensional similarity, and the effect on flight aerodynamics of that departure. *Trans. Am. Phil. Soc.* **65**, 1–67.

Hackett, S. J., Kimball, R. T., Reddy, S., Bowie, R. C. K., Braun, E. L., Braun, M. J., Chojnowski, J. L., Cox, W. A., Han, K.-L., Harshman, J., Huddleston, C. J., Marks, B. D., Miglia, K. J., Moore, W. S., Sheldon, F. H., Steadman, D. W., Witt, C. C. and Yuri, T. (2008). A phylogenomic study of birds reveals their evolutionary history. *Science* **320**, 1763–1768.

Hansen, T. F. and Martins, E. P. (1996). Translating between microevolutionary process and macroevolutionary patterns: the correlation structure of interspecific data. *Evolution* **50**, 1404–1417.

Hartman, F. A. (1961). *Locomotor Mechanisms of Birds*, Smithsonian Miscellaneous Collections, 143. Washington: Smithsonian Institution.

Harvey, P. H. and Pagel, M. D. (1991). *The Comparative Method in Evolutionary Biology*. Oxford: Oxford University Press.

Hedenström, A. (2003). Twenty-three testable predictions about bird flight. In: *Avian Migration* (ed. Berthold, P., Gwinner, E. and Sonnenschein, E.), pp. 563–582. Berlin: Springer-Verlag.

Hedenström, A., Johansson, C. and Spedding, G. R. (2009). Bird or bat: comparing airframe design and flight performance. *Bioinsp. Biomim.* **4**, 015001.

Hedenström, A. and Liechti, F. (2001). Field estimates of body drag coefficient on the basis of dives in passerine birds. *J. Exp. Biol.* **204**, 1167–1175.

Hedenström, A. and Møller, A. P. (1992). Morphological adaptations to song flight in passerine birds: a comparative study. *Proc. R. Soc. Lond. B.* **247**, 183–187.

Hertel, F. and Ballance, L. T. (1999). Wing ecomorphology of seabirds from Johnston Atoll. *Condor* **101**, 549–556.

Higgins, P. J., ed. (1999). *Handbook of Australian, New Zealand and Antarctic Birds. Volume 4: Parrots to Dollarbird*. Melbourne: Oxford University Press.

Higgins, P. J. and Davies, S. J. J. F., eds. (1996). *Handbook of Australian, New Zealand and Antarctic birds. Volume 3: Snipe to Pigeons*. Melbourne: Oxford University Press.

Higgins, P. J., Peter, J. M. and Cowling, S. J., eds. (2006). *Handbook of Australian, New Zealand and Antarctic Birds. Volume 7: Boatbill to Starlings*. Melbourne: Oxford University Press.

Higuera, C., Villaverde, J. R. B., Ross, J. and Morán, F. (2012). Multi-criteria optimization of regulation in metabolic networks. *PLoS ONE* **7**, e41122.

Horn, R. A. and Johnson, C. A. (2013). *Matrix analysis*, 2nd Ed. Cambridge: Cambridge University Press.

Huxley, J. S. and Teissier, G. (1936). Terminology of relative growth. *Nature* **137**, 780–781.

Iwasa, Y. (1988). Free fitness that always increases in evolution. *J. Theor. Biol.* **135**, 265–281.

Katz, J. and Plotkin, A. (2001). *Low-speed Aerodynamics*, 2nd Ed. Cambridge: Cambridge University Press.

Kelly, C. and Price, T. D. (2004). Comparative methods based on species mean values. *Math. Biosci.* **187**, 135–154.

Kerlinger, P. (1989). *Flight Strategies of Migrating Hawks*. Chicago: University of Chicago Press.

Kerney, K. P. (1972). A correction to "Lifting-line theory as a singular perturbation problem". *AIAA Journal* **10**, 1683–1684.

Kimura, M. (1958). On the change of population fitness by natural selection. *Heredity* **12**, 145–167.

Kingman, J. F. C. (1961). A mathematical problem in population genetics. *Math. Proc. Cambridge Phil. Soc.* **57**, 574–582.

Kleiber, M. (1932). Body size and metabolism. *Hilgardia* **6**, 315–353.

Kozłowski, J. and Konarzewski, M. (2004). Is West, Brown and Enquist's model of allometric scaling mathematically correct and biologically relevant? *Funct. Ecol.* **18**, 283–289.

Lauder, G. V. (2003). The intellectual challenge of biomechanics and evolution. In: *Vertebrate Biomechanics and Evolution* (ed. Bels, V. L., Gasc, J.-P. and Casinos, A.), pp. 319–324. Oxford: BIOS Scientific Publishers Ltd.

Lessard, S. (1997). Fisher's fundamental theorem of natural selection revisited. *Theor. Pop. Biol.* **52**, 119–136.

Lighthill, J. (1986). *An Informal Introduction to Theoretical Fluid Mechanics*, 2nd Ed. Oxford: Oxford University Press.

Lilienthal, O. (1889). *Der Vogelflug als Grundlage der Fliegerkunst*. Berlin: R. Gaertners Verlagsbuchhandlung.

Magnan, A. (1922). *Les caractéristiques des oiseaux suivant le mode de vol: leur application a la construction des avions*. Paris: Masson.

Maloly, G. M. O., Heglund, N. C., Prager, L. M., Cavagna, G. A. and Taylor, R. C. (1986). Energetic costs of carrying loads: have African women discovered an economic way? *Nature* **319**, 668–669.

Marchant, S. and Higgins, P. J., eds. (1990). *Handbook of Australian, New Zealand and Antarctic birds. Volume 1: Ratites to Ducks*. Melbourne: Oxford University Press.

Marchant, S. and Higgins, P. J., eds. (1993). *Handbook of Australian, New Zealand and Antarctic Birds. Volume 2: Raptors to Lapwings*. Melbourne: Oxford University Press.

Martins, E. P. and Hansen, T. F. (1997). Phylogenies and the comparative method: a general approach to incorporating phylogenetic information into the analysis of interspecific data. *Am. Nat.* **149**, 646–667.

Matthews, M. R. (2001). Methodology and politics in science: the fate of Huygens' 1673 proposal of the seconds pendulum as an international standard of length and some educational suggestions. In: *Science Education and Culture: The Role of History and Philosophy of Science* (ed. Mathews, M. R., Bevilacqua, F. and Giannetto, E.), pp. 293–309. Dordrecht: Kluwer.

Maybury, W. J., Rayner, J. M. V. and Couldrick, L. B. (2001). Lift generation by the avian tail. *Proc. R. Soc. Lond. B.* **268**, 1443–1448.

Maynard Smith, J., Buria, R., Kauffman, S., Alberch, P., Campbell, J., Goodwin, B., Lande, R., Raup, D. and Wolpert, L. (1985). Developmental constraints and evolution. *Q. Rev. Biol.* **60**, 265–287.

Mayr, G. (2011). Metaves, Mirandornithes, Strisores and other novelties — a critical review of the higher-level phylogeny of neornithine birds. *J. Zool. Syst. Evol. Res.* **49**, 58–76.

McCulloch, C. E., Searle, S. R. and Neuhaus, J. M. (2008). *Generalized, Linear, and Mixed Models*, 2nd Ed. New Jersey: Wiley.

McGahan, J. (1973). Gliding flight of the Andean condor in nature. *J. Exp. Biol.* **58**, 225–237.

McGhee, G. R. (2007). *The Geometry of Evolution: Adaptive Landscapes and Theoretical Morphospaces.* Cambridge: Cambridge University Press.

McMahon, T. A. and Bonner, J. T. (1983). *On Size and Life.* New York: Scientific American Library.

McMahon, T. A. and Cheng, G. C. (1990). The mechanics of running: how does stiffness couple with speed? *J. Biomech.* **23**, 65–78.

McMahon, T. A., Valiant, G. and Frederick, E. C. (1987). Groucho running. *J. Appl. Physiol.* **62**, 2326–2337.

Mendelssohn, J. M., Kemp, A. C., Biggs, H. C., Biggs, R. and Brown, C. J. (1989). Wing loadings and wing spans of 66 species of African raptors. *Ostrich* **60**, 35–42.

Müllenhoff, K. (1885). Die Grösse der Flugflächen. *Pflügers Archiv* **35**, 407–453.

Mustonen, V. and Lässig, M. (2009). Fitness flux and ubiquity of adaptive evolution. *PNAS* **107**, 4248–4253.

Nee, S. N. C., West, S. A. and Grafen, A. (2005). The illusion of invariant quantities in life histories. *Science* **309**, 1236–1239.

Norberg, R. Å. and Norberg, U. M. (1971). Take-off, landing, and flight speed during fishing flights of *Gavia stellata* (Pont.). *Ornis Scand.* **2**, 55–67.

Norberg, U. L. (1990). *Vertebrate Flight: Mechanics, Physiology, Morphology, Ecology and Evolution*, Zoophysiology, 27. Berlin: Springer-Verlag.

Norberg, U. M. (1986). Evolutionary convergence in foraging niche and flight morphology in insectivorous aerial-hawking birds and bats. *Ornis Scand.* **17**, 253–260.

Okasha, S. (2008). Fisher's Fundamental Theorem of Natural Selection: a philosophical analysis. *Brit. J. Phil. Sci.* **59**, 319–351.

Packard, G. C., Birchard, G. F. and Boardman, T. J. (2011). Fitting statistical models in bivariate allometry. *Biol. Rev.* **86**, 549–563.

Pagel, M. D. (1997). Inferring evolutionary processes from phylogenies. *Zool. Scripta* **26**, 331–348.

Paley, W. (1802). *Natural Theology, or, Evidences of the Existence and Attributes of the Deity, Collected from the Appearances of Nature*, 2nd Ed. London: R. Faulder.

Pennycuick, C. (1975). Mechanics of flight. In: *Avian Biology* (ed. Farner, D. S., King, J. R. and Parkes, K. C.), 5, pp. 1–75. London: Academic Press.

Pennycuick, C. J. (1971). Gliding flight of the white-backed vulture *Gyps africanus*. *J. Exp. Biol.* **55**, 13–38.

Pennycuick, C. J. (1978). Fifteen testable predictions about bird flight. *Oikos* **30**, 165–176.

Pennycuick, C. J. (1990). Predicting wingbeat frequency and wavelength of birds. *J. Exp. Biol.* **150**, 177–187.

Pennycuick, C. J. (1996). Wingbeat frequency of birds in steady cruising flight: new data and improved predictions. *J. Exp. Biol.* **199**, 1711–1726.

Pennycuick, C. J. (1999). *Measuring Birds' Wings for Flight Performance Calculations*, 2nd Ed. Bristol: Boundary Layer Publications.

Pennycuick, C. J. (2008). *Modelling the Flying Bird*. London: Academic Press.

Pigliucci, M. (2012). Landscapes, surfaces, and morphospaces: what are they good for? In: *The Adaptive Landscape in Evolutionary Biology* (ed. Svensson, E. I. and Calsbeek, R.), pp. 26–38. Oxford: Oxford University Press.

Poole, A., Stettenheim, P. R. and Gill, F., eds. (1992–2002). *The Birds of North America*. Washington, DC: American Ornithologists' Union.

Poole, E. L. (1938). Weights and wing areas in North American birds. *Auk* **55**, 511–517.

Pozo, C., Guillén-Gosálbez, G., Sorribas, A. and Jiménez, L. (2012). Identifying the preferred subset of enzymatic profiles in nonlinear kinetic metabolic models via multiobjective global optimization and Pareto filters. *PLoS ONE* **7**, e43487.

Price, G. R. (1970). Selection and covariance. *Nature* **227**, 520–521.

Price, G. R. (1972). Fisher's 'fundamental theorem' made clear. *Ann. Hum. Genet. Lond.* **36**, 129–140.

Raup, D. (1966). Geometric analysis of shell coiling: general problems. *J. Paleontol.* **40**, 1178–1190.

Rayner, J. M. V. (1985). Linear relations in biomechanics: the statistics of scaling functions. *J. Zool. Lond. (A)* **206**, 415–439.

Rayner, J. M. V. (1988). Form and function in avian flight. In: *Current Ornithology* (ed. Johnston, R. F.), vol. 5, pp. 1–66. New York: Plenum Press.

Rencher, A. C. and Schaalje, G. B. (2008). *Linear Models in Statistics*. New Jersey: Wiley.

Ridley, M. (1983). *The Explanation of Organic Diversity: The Comparative Method and Adaptations for Mating*. Oxford: Clarendon Press.

Riska, B. (1991). Regression models in evolutionary allometry. *Am. Nat.* **138**, 283–299.

Rohlf, F. J. (2001). Comparative methods for the analysis of continuous variables: geometric interpretations. *Evolution* **55**, 2143–2160.

Rohlf, F. J. (2006). A comment on phylogenetic correction. *Evolution* **60**, 1509–1515.

Rosenzweig, M. L. (1978). Competitive speciation. *Biol. J. Linn. Soc.* **10**, 275–289.

Sachs, G., Traugott, J., Nesterova, A. P., Dell'Omo, G., Kümmeth, F., Heidrich, W., Vyssotski, A. L. and Bonadonna, F. (2012). Flying at no mechanical energy cost: disclosing the secret of wandering albatrosses. *PLoS ONE* **7**, e41449.

Savage, V. M., Gillooly, J. F., Brown, J. H., West, G. B. and Charnov, E. L. (2004a). Effects of body size and temperature on population growth. *Am. Nat.* **163**, 429–441.

Savage, V. M., Gillooly, J. F., Woodruff, W. H., West, G. B., Allen, A. P., Enquist, B. J. and Brown, J. H. (2004b). The predominance of quarter-power scaling in biology. *Funct. Ecol.* **18**, 257–282.

Schmidt-Nielsen, K. (1984). *Scaling: Why is Animal Size so Important?* Cambridge: Cambridge University Press.

Schuetz, R., Zamboni, N., Zampieri, M., Heinemann, M. and Sauer, U. (2011). Multidimensional optimality of microbial metabolism. *Science* **336**, 601–604.

Shoval, O., Sheftel, H., Shinar, G., Hart, Y., Ramote, O., Mayo, A., Dekel, A., Kavanagh, K. and Alon, U. (2012). Evolutionary trade-offs, Pareto optimality, and the geometry of phenotype space. *Science* **336**, 1157–1160.

Shoval, O., Sheftel, H., Shinar, G., Hart, Y., Ramote, O., Mayo, A., Dekel, A., Kavanagh, K. and Alon, U. (2013). Response to comment on "Evolutionary trade-offs, Pareto optimality, and the geometry of phenotype space". *Science* **339**, 757.

Sieg, A. E., O'Connor, M. P., McNair, J. N., Grant, B. W., Agosta, S. J. and Dunham, A. E. (2009). Mammalian metabolic allometry: do intraspecific variation, phylogeny, and regression models matter? *Am. Nat.* **174**, 720–733.

Simons, E. L. (2010). Forelimb skeletal morphology and flight mode evolution in pelecaniform birds. *Zoology* **113**, 39–46.

Simpson, G. G. (1944). *Tempo and Mode in Evolution*. New York: Columbia University Press.

Sober, E. (2008). *Evidence and Evolution: The Logic Behind the Science*. Cambridge.: Cambridge University Press.

Solari, M. E. (1969). The "maximum likelihood solution" of the problem of estimating a linear functional relationship. *J. R. Stat. Soc. B* **31**, 372–375.

Spear, L. B. and Ainley, D. G. (1997). Flight behaviour of seabirds in relation to wind direction and wing morphology. *Ibis* **139**, 22–233.

Sprent, P. and Dolby, G. R. (1980). The geometric mean functional relationship. *Biometrics* **36**, 547–550.

Strawson, P. F. (1950). On referring. *Mind* **235**, 320–344.

Swartz, S. M. (1989). Pendular mechanics and the kinematics and energetics of brachiating locomotion. *Int. J. Primatol.* **10**, 387–418.

Taylor, G. K., Nudds, R. L. and Thomas, A. L. R. (2003). Flying and swimming animals cruise at a Strouhal number tuned for high power efficiency. *Nature* **425**, 707–711.

Taylor, G. K., Triantafyllou, M. S. and Tropea, C., eds. (2010). *Animal Locomotion: The Physics of Flying; the Hydrodynamics of Swimming*. Berlin: Springer-Verlag.

Thomas, A. L. R. (1993). On the aerodynamics of birds' tails. *Phil. Trans. R. Soc. Lond. B* **340**, 361–380.

Thomas, A. L. R. and Taylor, G. K. (2001). Animal flight dynamics I. Stability in gliding flight. *J. Theor. Biol.* **212**, 399–424.

Thomas, F. (1999). *Fundamentals of Sailplane Design*, 3rd Ed. College Park MA: College Park Press.

Thompson, D. W. (1917). *On Growth and Form*. Cambridge: Cambridge University Press.

Tinbergen, N. (1963). On aims and methods of ethology. *Z. Tierpsychol.* **20**, 410–433.

Tobalske, B. W. (1996). Scaling of muscle composition, wing morphology, and intermittent flight behavior in woodpeckers. *Auk* **113**, 151–177.

Tobalske, B. W. (1999). Kinematics of flap-bounding flight in the zebra finch over a wide range of speeds. *J. Exp. Biol.* **202**, 1725–1739.

Tritton, D. J. (1988). *Physical Fluid Dynamics*. Oxford: Oxford University Press.

Tucker, V. A., Cade, T. J. and Tucker, A. E. (1998). Diving speeds and angles of a gyrfalcon (*Falco rusticolus*). *J. Exp. Biol.* **201**, 2061–2070.

Usherwood, J. R. (2005). Why not walk faster? *Biol. Lett.* **1**, 338–341.

Van Dyke, M. (1964). *Perturbation Methods in Fluid Mechanics*. New York: Academic Press.

Videler, J. J. (2005). *Avian Flight*. Oxford: Oxford University Press.

Vogel, S. (2003). *Comparative Biomechanics*. Princeton, NJ: Princeton University Press.

Wang, X., Nudds, R. L. and Dyke, G. J. (2011). The primary feather lengths of early birds with respect to avian wing shape evolution. *J. Evol. Biol.* **24**, 1226–1231.

Warham, J. (1977). Wing loadings, wing shapes, and flight capabilities of Procellariiformes. *N.Z. J. Zool.* **4**, 73–83.

Warmflash, A., Francois, P. and Siggia, E. D. (2012). Pareto evolution of gene networks: an algorithm to optimize multiple fitness objectives. *Phys. Biol.* **9**, 056001.

Warton, D. I., Wright, I. J., Falster, D. S. and Westoby, M. (2006). Bivariate line-fitting methods for allometry. *Biol. Rev.* **81**, 259–291.

Wenham, F. (1866). *Aerial Locomotion*, Aeronautical Classics, 2. London: Royal Aeronautical Society.

West, G. B., Brown, J. H. and Enquist, B. J. (1997). A general model for the origin of allometric scaling laws in biology. *Science* **276**, 122–126.

West, G. B., Brown, J. H. and Enquist, B. J. (2001). A general model for ontogenetic growth. *Nature* **413**, 628–631.

West, G. B., Woodruff, W. H. and Brown, J. H. (2002). Allometric scaling of metabolic rate from molecules and mitochondria to cells and mammals. *PNAS* **99**, 2473–2478.

White, C. R. (2011). Allometric estimation of metabolic rates in animals. *Comp. Biochem. Physiol. A* **158**, 346–357.

White, C. R. and Seymour, R. S. (2003). Mammalian basal metabolic rate is proportional to body mass 2/3. *PNAS* **100**, 4046–4049.

Wright, S. (1932). The roles of mutation, inbreeding, crossbreeding and selection in evolution. In: *Proceedings of the Sixth International Congress of Genetics*, pp. 356–366. Brooklyn, NY: Brooklyn Botanic Garden.

Wright, S. (1988). Surfaces of selective value revisited. *Am. Nat.* **131**, 115–123.

Zhang, F., Xu, L., Zhang, K., Wang, E. and Wang, J. (2012). The potential and flux landscape theory of evolution. *J. Chem. Phys.* **137**, 065102.

Index

Bold entries denote pages at which technical terms are defined. Entries in *italics* refer to figures or tables.